ALGEBRA EXAMPLES

POWERS AND LOGARITHMS 1

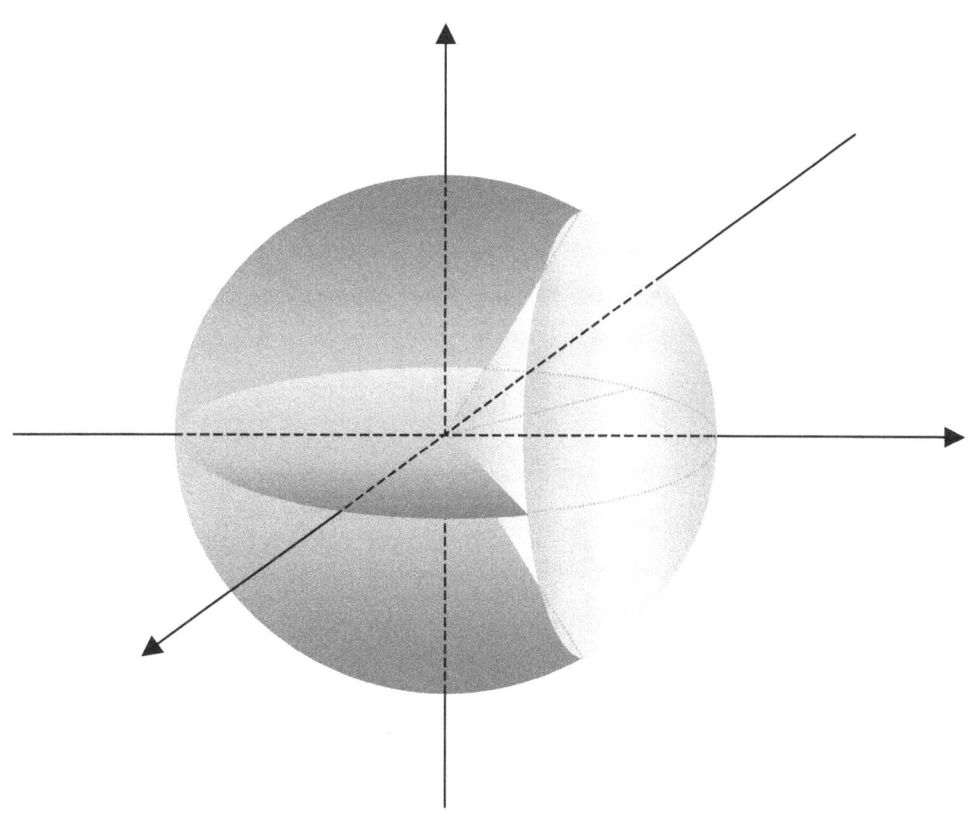

Seong R. KIM

Dear students:

Students need the best teacher, so you need examples, because examples are the best teacher. All the examples here are fully worked, and explain **how** the basic and essential tools in math are made, together with **what** they are, **how** they work, and **how** to work with them. Such tools include numbers, formulas, identities, equations, laws, etc.

Examples here begin with easy ones, of course. Covering every meter and yard properly, we can cover thousands of miles and kilometers. And it is particularly the case in math.

Of those examples therefore, some might even look too easy for you. It's not that easy though, to come up with those examples. Anyways, the bigger and the taller the tree, the deeper and the stronger the root.

Doing math, we work with ideas and run ideas, because every thing in math is an idea. A number is an idea, for instance, and the same is true for a line or circle, too. And putting ideas together, we build another, which becomes the base or an element of another, and each is connected. And that's the way your math grows. So you get to build a circuit, and sometimes, need to fill the gap or repair the circuit so that you get the sense of it.

So your calculation runs properly, and you get the problem solved.

The examples have been made and arranged so that they get tougher (or sometimes easier for some reason) as you proceed with them. In particular, similar examples with some variations are strategically repeated so that you can get the ideas or the tools tricky or complicated, and can get them mastered.

This book is however, nothing but a bunch of examples until you get it powered. How then, to get it powered, and make it run and work for you?

Just read it, and then, do each example in writing. And it is important to note that you do it in **your** writing. Just watching someone doing it, you just only feel that you can do it. If you do it, you can do it, but if you don't, we can hardly. It's a cliché, of course, but is always true that knowing is one thing and doing is another.

I've been helping students grow, take care of, and run their own math. The area covers algebra and geometry for high school or college students, and is especially for equations (for unknowns or curves), functions, and their graphs, which are the basic elements in calculus, which's been the core of my interest from my early age in high school.

Of my students, some are quite poor in math, and thus, are afraid of or hate math, some require special education because of exceptional intelligence, some are smart enough, some are naïve and diligent, some are clever but lazy, and most behave in general. All the students are badly after though, one thing in common: a strong and secure math skill. It is of course, the prime objective of my work, and I'm always happy to and eager to help them achieve it. The problem was however, that many of them wanted it to be purchased. And the question is, can we buy it?

We can buy the means, of course. And a solid math skill is feasible, too. We know however, we can't buy love, and the same is true for the math skill, too. It's not what we can buy or sell, and not what we can give or take. It is however, what we can grow, and need to grow. Your math grows as much as you grow and take care of it. So does mine.

What math then, do students most often do or use in high schools or colleges?

It is algebra and geometry. What algebra though?

Elementary algebra, of course
Doing the algebra, we work with numbers (many in kinds), constants, variables, ratios, rates, expressions, equations, inequalities, functions, identities, formulas, laws, etc., together with signs and symbols. And if we want to do algebra properly, we want to know their natures and how they mingle with each other.

So studying math ideas or tools, you want to know **what** they are, **how** they work, and **how** to work with them or **what** to do with them.   What then, about the geometry?

Basically, the geometry has much to do with shapes, positions, and angles. The shapes begin with triangles and circles, and move on to rectangles, squares, parallelograms or rhombuses, trapezoids, tetragons, other polygons, polyhedrons, etc.

Doing the geometry, too, though, we need to do the algebra stated above. So it is analytic geometry, often called coordinate geometry, too. And doing it, we can specify positions using coordinates. So in the geometry, basically, we work with graphs. Putting a math idea in a graph, we can not only effectively think about it but actually see it, too, and therefore, can efficiently work with it. What idea then, is it?

The idea begins with a point, line, parabola, circle, ellipse, and hyperbola, called a conic section or basic curve, and then, moves on to other curves, planes, surfaces, volumes, and other objects in various dimensional spaces, together with vectors.

And using an angle, we can specify an amount of turn or change in direction.

So learning, using, or applying those ideas or math tools, we get to solve problems.

And this book can help. It can help learn them, and use them so that you can navigate to find solutions to problems. And in particular, it can help come up with answers to those **what**s and **how**s stated above. So it can help you grow and run your own math, and thus, can help achieve your solid math skill.

It is however, not a magic book giving you a math skill of high caliber overnight. And it can have many mistakes, too. There is no magic, and math is full of facts and ideas. And it is after all, not me and not your teacher but you who put together some of those facts and ideas, and understand it. Putting facts and ideas together, understanding it, and taking care of what you have learned, you grow your math. And this book can help.

This is a book of examples designed to help you grow your math, and assumes that you are a real beginner. This book requires though, time and effort, the amount of which need to be substantial, too, but will be worth it. That's because you want a substantial achievement, and will get it. And probably, you will get to see this book helping you get there much faster than expected. And then, you will get to see the way math runs.

In math, everything is an idea. So is a problem. And solving it, we put it many different ways. For instance, while expanding or reducing it, or modifying or converting it, we keep searching for the solution, approaching the solution, and eventually, can get there. So don't look for the solution outside the problem. The solution is inside the problem if the problem is properly made.

If it is not, no solution is the solution. And in fact, it is often the case a problem itself is the solution. We can put a problem in many different ways, and eventually, can end up with the solution.   How come then, is the solution no other than the problem?

For instance, the solution to $3232 \div 101$ is 32.   And we can put it this way:

$$3232 \div 101 = \frac{3232}{101} = \frac{32 \times 101}{101} = \frac{32}{1} = 32 \implies 3232 \div 101 = 32.$$

And we can get this, too: $32 \implies 3232 \div 101$.   How?

$$32 = \frac{32}{1} = \frac{32 \times 101}{101} = \frac{3232}{101} = 3232/101 = 3232 \div 101.   \text{Too easy?}$$

For another instance, the solution to $ax^2 + bx + c = 0$ is: $x = \frac{-b \pm \sqrt{b^2 - 4ac}}{2a}$, which is called the quadratic formula.   How come then, is the solution no other than the problem?

We can put it this way:

$$x = \frac{-b \pm \sqrt{b^2 - 4ac}}{2a} \implies 2ax = -b \pm \sqrt{b^2 - 4ac} \implies 2ax + b = \pm\sqrt{b^2 - 4ac}$$

$$\implies (2ax + b)^2 = b^2 - 4ac \implies 4a^2x^2 + 4abx + b^2 = b^2 - 4ac$$

$$\implies 4a^2x^2 + 4abx = -4ac \implies ax^2 + bx = -c \implies ax^2 + bx + c = 0.$$

And we can get this, too: $ax^2 + bx + c = 0 \implies x = \frac{-b \pm \sqrt{b^2 - 4ac}}{2a}$.   How?

$$ax^2 + bx + c = a(x^2 + \tfrac{b}{a}x) + c = a(x^2 + \tfrac{b}{a}x + \tfrac{b^2}{4a^2} - \tfrac{b^2}{4a^2}) + c = a(x^2 + \tfrac{b}{a}x + \tfrac{b^2}{4a^2}) - \tfrac{b^2}{4a} + c$$

$$= a(x + \tfrac{b}{2a})^2 - \tfrac{b^2 - 4ac}{4a} = 0 \implies a(x + \tfrac{b}{2a})^2 = \tfrac{b^2 - 4ac}{4a} \implies (x + \tfrac{b}{2a})^2 = \tfrac{b^2 - 4ac}{4a^2} \implies x + \tfrac{b}{2a} = \pm\sqrt{\tfrac{b^2 - 4ac}{4a^2}}$$

$$\implies x = -\tfrac{b}{2a} \pm \tfrac{\sqrt{b^2 - 4ac}}{2a} = \tfrac{-b \pm \sqrt{b^2 - 4ac}}{2a} \implies x = \tfrac{-b \pm \sqrt{b^2 - 4ac}}{2a}.$$

And we call the set of processes above, algebra.

So if a problem is well defined, that is, if it makes sense, we should be able to get it solved the way below:

A problem $\Rightarrow$ **...** $\Rightarrow$ **...** $\Rightarrow$ **the solution**, and thus: **the problem** $\Rightarrow$ **the solution**.

So solving a problem, we put it many different ways so that we can get to the solution.

And that's the way, math runs.

May your math run very well.

Seong R. Kim

B.S. Math. Michigan Tech. Univ.   M.S. Math. Rensselaer Polytechnic Institute

Notes:

This book is about math ideas called powers. Why powers though?

They have numbers called exponents, and are in fact, for your algebra. Doing algebra, you need to know what powers and exponents are about, how they work so that you can work with them. Why algebra though?

It's simply because you need to *solve problems*. Algebra connects problems to solutions. Algebra only can in fact, actually get you the ones you are always busy finding when taking exams, tests, and quizzes, and doing homework, too, of course. You've got to do algebra to get the very one you want so that you can put it down on *your answer sheet*. With algebra skill, together with your creativity, you can actually solve problems.

And this book is in fact, for your skill of algebra, and you will grow it through examples. Some examples may look too easy or too hard. It all depends on your skill of algebra. Whatever your skill may be though, you can grow yours if you follow the steps in each example. Each is detailed so that you can learn those tools fast, and increase your caliber quickly as well as properly.

And this book explains what powers and exponents are about and how to manipulate powers, that is, how to change or alter, convert, or modify those math expressions so that you can come up with the ones that you need. The ones are solutions, of course. And that's what this book is about.

This book does not just explain though. But it helps follow steps to the solutions, too, and thus, helps you do calculations with powers and exponents so that you can actually do the calculation work doing those manipulations above.

With strong algebra, you can learn things in math fast, and can do problems very well, too, of course. And this book is about exponential algebra, which is on expressions with powers. And there are two more books related to this book, and the two are as follows:

ALGEBRA EXAMPLES POWERS AND LOGARITHMS 2

ALGEBRA EXAMPLES POWERS AND LOGARITHMS 3

And also, all the basics on powers and logarithms and all the ideas contained in this book and the two books above are covered in one book, too. And the book is as follows:

ALGEBRA EXAMPLES POWERS AND LOGARITHMS

So either way, the books will get you not only powers and logarithms but enhancement of your algebra, too. You will thus, soon be able to control powers and logarithms, that is, change or alter, convert, or modify those math expressions so that you can get to the solutions fast. And you will learn them all through examples detailed so that your math can run not only properly but fast enough, too.

Contents

In POWERS AND LOGARITHMS 1

The Preview of the Contents

In POWERS AND LOGRAITHMS 2

The Preview of the Contents

In POWERS AND LOGARITHMS 3

$$(x + y)^2 = x^2 + 2xy + y^2. \qquad\qquad (x + y)^3 = x^3 + 3x^2y + 3xy^2 + y^3.$$

$$(x + y)(x - y) = x^2 - y^2. \qquad\qquad (x + y)(x^2 - xy + y^2) = x^3 + y^3.$$

$$(x^2 + xy + y^2)(x^2 - xy + y^2) = x^4 + x^2y^2 + y^4.$$

$$(x + a)(x + b) = x^2 + (a + b)x + ab. \qquad (ax + b)(cx + d) = acx^2 + (ad + bc)x + bd.$$

$$(x + a)(x + b)(x + c) = x^3 + (a + b + c)x^2 + (ac + bc + ca)x + abc.$$

$$(a + b + c)^2 = a^2 + b^2 + c^2 + 2(ab + bc + ca).$$

$$(a + b + c)(a^2 + b^2 + c^2 - ab - bc - ca) = a^3 + b^3 + c^3 - 3abc.$$

Suppose both a and $b \neq 0$, and both m and n are integers. Then, we get:

0. $a^m a^n = a^{m+n}$ **1.** $a^m / a^n = \dfrac{a^m}{a^n} = a^{m-n}$ **2.** $(a^m)^n = a^{mn}$

3. $(ab)^n = a^n b^n$ **4.** $(a/b)^n = \left(\dfrac{a}{b}\right)^n = a^n / b^n = \dfrac{a^n}{b^n}$

Suppose both a and $b > 0$, and m and n both are integers nonzero. Then, we get:

0.1. $a^{\frac{1}{n}} b^{\frac{1}{n}} = (ab)^{\frac{1}{n}}$.   **1.1.** $\dfrac{a^{\frac{1}{n}}}{b^{\frac{1}{n}}} = \left(\dfrac{a}{b}\right)^{\frac{1}{n}}$.   **2.1.** $(a^{\frac{1}{n}})^m = (a^m)^{\frac{1}{n}}$.

3.1. $(a^{\frac{1}{n}})^{\frac{1}{m}} = a^{\frac{1}{mn}} = (a^{\frac{1}{m}})^{\frac{1}{n}}$.   **3.2.** $(a^{mp})^{\frac{1}{np}} = (a^m)^{\frac{1}{n}}$, where p is a nonzero integer.

1. Suppose M, N, and $b > 0$, but $b \neq 1$, and we have: $A = \log_b M$, and $B = \log_b N$. Then, we get: $A - B = \log_b M - \log_b N = \log_b \frac{M}{N}$.

2. Suppose that M and $b > 0$, but $b \neq 1$, and that we have: $E = \log_b M$. Then, we get: $PE = P \log_b M = \log_b M^P$.

3. Suppose that a, b, C, and $D > 0$, but a and $b \neq 1$, and that we have: $\log_a C = \log_b D$. Then, we get: $\log_a C = \log_b D = \log_{ab} CD$.

4. Suppose that a, b, C, and $D > 0$, but a and $b \neq 1$, and that we have: $\log_a C = \log_b D$. Then, we get: $\log_a C = \log_b D = \log_{\frac{a}{b}} \frac{C}{D} = \log_{\frac{b}{a}} \frac{D}{C}$.

5. $\log_b b = 1$, and $\log_b 1 = 0$.   **6.** $\log_b A = \dfrac{\log_c A}{\log_c b}$.

7. $\log_b A = \dfrac{1}{\log_A b}$.

Note:

The drawings or graphs in this book are not exact, and are approximate or conceptual ones.

$\in$	"$a \in B$" means that a belongs to B. "p, q, and $r \in W$" means that p, q, and r belong to W.
$\Rightarrow$	"$A \Rightarrow B$." means that A implies B.
$\equiv$	$A \equiv B$ means that A and B are identical to each other.
$\neq$	$A \neq B$ means that A is not equal to B.
$|A|$	The magnitude of A. For instance, $|\text{-}1| = |1| = 1$.
$\therefore$	Therefore
$\Leftrightarrow$	"$A \Leftrightarrow B$" means "If A then B." and "If B then A." We can read $A \Leftrightarrow B$ as "A if and only if B." In such a case, we can say that $A = B$.
Δx and Δy	Suppose that (x_1, y_1) and (x_2, y_2) are two points in the x-y plane. Then, we get either of the two below. $\Delta x = x_2 - x_1$, and $\Delta y = y_2 - y_1$. $\Delta x = x_1 - x_2$, and $\Delta y = y_1 - y_2$.

Distance Formula

Suppose that d is the distance between two points (x_1, y_1) and (x_2, y_2) in the x-y plane. Then, we get $d^2 = (\Delta x)^2 + (\Delta y)^2$.

0. What is a power?

It is a tool, and in fact, is a power tool, has quite a bit of power, and gets powered by numbers, together with your reasoning. Being a power user of such a tool, you can grab and handle readily many values, excessively big as 5233476330273605372135115521 or unreasonably small as 0.00000000000000022876792454961. And also, taking care of ordinary values as 64, 243, 625, 1024, and such, you can effectively handle them, too, using the tool called a power, of course.   What then, is the tool?

When adding together many of the same numbers, we don't actually do additions so many times, do we? We just multiply, of course, one of those many by the number of those, and get the product, which is the sum.

So for instance, adding together five **3**s, we don't normally do: **3 + 3 + 3 + 3 + 3**, but just do: **3·5**, that is, **3 × 5**, and get **15**, which is the sum as well as the product of 3 and 5.

Next, what if we want to multiply 1 by 3 many times, and specify the product?

If we just want to multiply 1 by 3 twice, we just do: 1·3·3, and get 9, which is the product.
What if however, we want to multiply 1 by 3 five hundred times, and specify the product?

Let alone specifying such a product, we are probably not going to actually do multiplications so many times, are we?

Suppose however, we actually did do multiplications so many times, and somehow did find the product. Then, we would get a big number, which is a long number.

The number is the product of five hundred **3**s, and has a huge number of digits. In fact, it has a whopping array of 239 digits. So it's an unreasonably long number with a huge value.   What if however, we've got to specify such a huge value anyway?

We don't actually specify such a long number digit by digit. We can quickly put such an enormous value in a special form or manner. Using a special notation, we can effectively specify a humongous value or a number with a gigantic value.

A number indicates a value.
And we can put a value in various forms. We have a variety of ways or methods, where we can indicate the same value. Depending on the purpose, we can show a value in a particular form. For instance, we can put **8** in such many ways as follows:

$$1 \times 8, \; 8 \cdot 1, \; 4 \cdot 2, \; 4 \cdot (3-1), \; 1+2+3+2, \; 3+1+4, \; 1+3-3+7, \; 9-1, \; \tfrac{16}{2}, \; \tfrac{3 \cdot 8}{3}, \text{etc.}$$

Quite often, we need to show values with unduly large or extremely small magnitudes, but want to express them in a succinct manner. If for instance, we have to put in writing values like 100000000000000000 and 0.000000000001, they not only take up too much space, but are hard to read, too.   What then, do we do?

In such a case, we can use a tool, called a *power*. And the tool takes two numbers, one is called an *exponent*, and the other is called a *base*. So we get the tool powered by two numbers called an *exponent* and a *base*.

Suppose for instance, a value is a product we get multiplying 1 by 7 a particular number of times. Then, we can specify the value using a power. Then, the particular number is called the *exponent*, and 7 is called the *base*.

So for instance, multiplying 1 by 9 twenty eight times, and expressing the product using a power, we use 9 as the base, and use 28 as the exponent.

So expressing 1·9·9·9·9·9·9·9·9·9·9·9·9·9·9·9·9·9·9·9·9·9·9·9·9·9·9·9·9 which is 523347633027360537213511521, we can use a power, where the base is 9, and the exponent is 28.   How then, can we specify a power?

Specifying a power, we put it in terms of two numbers, one is called a base, and the other is called an exponent. So putting a value in terms of a base and an exponent, we get a number called a power. A power is therefore, a structured number composed of two parts, one is a base, and the other is an exponent.

And **in this book**, such a structured number is said to be in *power notation.*
Using power notation, we put an exponent at the upper right hand corner of a base.
In other words, we use an exponent as a superscript.

So for instance, using power notation, we can put 8 in 2^3, and put 1,000,000,000 in 10^9.

And 2^3 is called a power, 2 is called the base, and 3 is called the exponent. And the same is true for the power 10^9, too. So 10 is the base of the power 10^9, and 9 is the exponent.

Putting thus, a value or number in power notation, we make a power, which is a number made of a base and an exponent.

And also, calling a power with a particular base, we call it a power of the particular base.

So for instance, 10^9 can be called a power of 10, and specifically, it is called the *ninth power* of 10. And also, we can call it differently, too, and will cover shortly how to do so.

And for another instance, we can simply put 523347633027360537213511521 in 9^{28}, which is a power of 9, and specifically, is called the twenty eighth power of 9. And thus, we can put such a huge value in a succinct manner.

What then, about small numbers as 0.0000001?

In such a case, we can use a *negative* number as the *exponent.*

A value can be a reciprocal of a large number, so such a value is very small, and we can get such a number dividing 1 by a particular number many times. For instance, it can be 1/23/23/23/23/23/23, which equals $\frac{1}{148035889}$, which we get dividing 1 by 23 six times.

And expressing a number we get doing divisions by the same number repeatedly, we can use power notation, too. Then, the base is the divisor, which is 23 in the case above, but the exponent is a negative integer, because a division is opposite of a multiplication. So –6 is the exponent, because the division by 23 happens 6 times.

Using thus, power notation, we put $\frac{1}{148035889}$ in 23^{-6}, which is a power of 23, and specifically, is called the *negative sixth* power of 23.

And we put 0.0000001 in 10^{-7}, because 0.0000001 = 1/10/10/10/10/10/10/10. And 10^{-7} is a power of 10, and is called the negative seventh power of 10.

And multiplying for instance, 1 by b twice, and specifying the product, we can do this: $1{\cdot}b{\cdot}b$, and then, can put the product this way: bb. Normally though, we put it this way: b^2. And dividing 1 by b twice, we put the result this way: b^{-2}. What if however, we want to indicate a value we get multiplying (or dividing) 1 by b an arbitrary number of times?

We can indicate the value by: b^n where n is an integer. So the superscript n indicates the number of the multiplications made. And the same is true for divisions, too, of course.

So in the power b^n where n is an integer, n indicates the number of times we multiply or divide 1 by b. For instance, $1{\cdot}b{\cdot}b{\cdot}b{\cdot}b{\cdot}b = 1bbbbb = b^5$, and $1/b/b/b/b = b^{-4}$. And of course, in cases of divisions, the base b is not 0.

What do we get though, if we put 0 into n in the power b^n?

Then, we do neither multiply nor divide 1 by the base b. That is, we do nothing to 1. In other words, we leave 1 alone. So we get: $b^0 = 1$, and for instance, $2^0 = 3^0 = 4^0 = 1$.

What if n is 1 in b^n?

Then, we multiply 1 by b once, and therefore, we get a b. So we get: $b^1 = b$.

And thus, we can put a value in a form of a power, which takes a form of b^n, where b is called a base, and n is called an exponent.   How can we read a power though?

We can read b^n as 'b raised to the nth power' or 'the nth power of b'. And briefly, we can read it as 'b to the nth'. Even more briefly, we can just read it as 'b to the n'.

For instance, we can read b^7 as 'b raised to the seventh power', 'the seventh power of b', or 'b to the seventh'. And more briefly, we can read it as 'b to the seven', too.

Usually though, we read b^2 differently, and often read it as 'b squared', instead of reading it as 'b raised to the second power', 'the second power of b', or 'b to the second'.

And the same is true for b^3, too. So b^3 is often read as 'b cubed'. And of course, we can read it as 'b to the third', too.   How then, about b^{-3}?

It can be read as 'b to the negative (minus) third', 'the negative third power of b', or just can be read briefly as 'b to the negative three'.
And in fact, we can use any real number as an exponent if the base is positive.

So for instance, we can read:

$b^{0.3}$ as 'b to the 0.3', b^{π} as 'b to the pi', and $b^{\sqrt{2}}$ as 'b to the square root of 2'.

0.3^3 as 0.3 cubed, and of course, we can read it as 0.3 to the third, too.

$b^{\frac{1}{n}}$ as 'b to the 1 over n', $b^{-\frac{y}{x}}$ as 'b to the negative y over x', and $b^{\frac{3}{\pi}}$ as 'b to the 3 over pi'.

$b^{-\frac{2}{3}}$ as 'b to the negative 2 over 3', and also, as 'b to the negative two thirds'.

$b^{f(2)}$ as 'b to the $f(1)$', and $b^{g(x)}$ as 'b to the $g(x)$'.

b^{x^2} as 'b to the x squared' or 'b to the x to the second', and $b^{x^{\frac{1}{3}}}$ as 'b to the x to the third'.

Examples in Powers

Note that as the multiplication operator, we often use a dot · instead of 'x'.

0. Expand (find the values of) the powers below:

For example, $2^2 = 2 \times 2 = 2 \cdot 2 = 4$.

$2^1,$ $2^3,$ $3^3,$ $4^3,$ $5^3,$ $6^3,$ $7^3,$ $8^3,$ $9^3,$ $1^{100}.$

1. Expand the powers below:

For example, $2^2 = 2 \times 2 = 2 \cdot 2 = 4$.

$2^{-1},$ $2^{-2},$ $3^{-3},$ $4^{-3},$ $5^{-3},$ $6^{-3},$ $7^{-3},$ $8^{-3},$ $9^{-3},$ $1^{-100}.$

2. Expand (find the values of) the expressions below:

For example, $2^2 3^2 = 2 \times 2 \times 3 \times 3 = 2 \cdot 2 \cdot 3 \cdot 3 = 4 \cdot 9 = 36$.

$2^2 3^3,$ $2^3 10^3,$ $2^3 7^2,$ $2^3 5^2 9^2.$

3. Expand the expressions below:

For example, $2^{-2} 3^{-2} = (1/2/2)(1/3/3) = (1/4) \cdot (1/9) = 1/36$.

$2^{-2} 3^{-3},$ $2^{-3} 10^{-3},$ $2^{-3} 7^{-2},$ $2^{-3} 5^{-2} 9^{-2}.$

Suggestions or Solutions
To the Examples in Powers

0.

$2^1 = 2$

$2^3 = 2 \times 2 \times 2 = 8$

$3^3 = 3 \times 3 \times 3 = 27$

$4^3 = 4 \times 4 \times 4 = 16 \times 4 = 40 + 24 = 64$

$5^3 = 5 \times 5 \times 5 = 25 \times 5 = 100 + 25 = 125$

$6^3 = 6 \times 6 \times 6 = 36 \times 6 = 180 + 36 = 216$

$7^3 = 7 \times 7 \times 7 = 49 \times 7 = 280 + 63 = 343$

$8^3 = 8 \times 8 \times 8 = 64 \times 8 = 480 + 32 = 512$

$9^3 = 9 \times 9 \times 9 = 81 \times 9 = 720 + 9 = 729$

$1^{100} = 1$

1.

$2^{-1} = 1/2$

$2^{-3} = 1/2/2/2 = 1/2/4 = 1/8$, which is $1/2^3$

$3^{-3} = 1/3/3/3 = 1/3/9 = 1/27$, which is $1/3^3$

$4^{-3} = 1/4/4/4 = 1/16/4 = 1/64$, which is $1/4^3$

$5^{-3} = 1/5/5/5 = 1/25/5 = 1/125$, which is $1/5^3$

$6^{-3} = 1/6/6/6 = 1/36/6 = 1/216$, which is $1/6^3$

$7^{-3} = 1/7/7/7 = 1/49/7 = 1/343$, which is $1/7^3$

$8^{-3} = 1/8/8/8 = 1/64/8 = 1/512$, which is $1/8^3$

$9^{-3} = 1/9/9/9 = 1/81/9 = 1/729$, which is $1/9^3$

$1^{-100} = 1/1^{100} = 1/1 = 1$

2.

$2^2 3^3 = 2 \times 2 \times 3 \times 3 \times 3 = 4 \times 27 = 80 + 28 = 108$

$2^3 10^3 = 8 \times 10 \times 10 \times 10 = 8000$

$2^3 7^2 = 8 \times 49 = 320 + 72 = 392$

$2^3 5^2 9^2 = 8 \times 25 \times 81 = (160 + 40) \times 81 = 200 \times 81 = 16000 + 200 = 16200$

3.

$2^{-2} 3^{-3} = (1/2/2)(1/3/3/3) = (1/4)(1/27) = 1/108$, which is $1/(2^2 3^3) = \frac{1}{2^2 3^3}$

$2^{-3} 10^{-3} = (1/8)(1/10/10/10) = 1/8000$, which is $1/(2^3 10^3) = \frac{1}{2^3 10^3}$

$2^{-3} 7^{-2} = (1/8)(1/49) = 1/392$, which is $1/(2^3 7^2) = \frac{1}{2^3 7^2}$

$2^{-3}5^{-2}9^{-2}$ = (1/8)(1/25)(1/81) = 1/(8·25·81) = 1/{(160 + 40) x 81} = 1/(200 x 81)

= 1/(16000 + 200) = 1/16200, which is $1/(2^3 5^2 9^2)$ = $\frac{1}{2^3 5^2 9^2}$

1. Exponential Identities 1

Doing math or taking courses in math, we can't do much without doing algebra. And doing algebra, we get to work with many tools as formulas, identities, theorems, etc. Or rather, whatever math we do, we have to do algebra, and always need to use such tools.

And also, doing algebra, we often get to work with powers. Or rather, we always have to work with powers. And doing algebra with powers, we do exponential algebra, and get to work with exponents and bases, of which the powers are made.

And doing exponential algebra, we can hardly do much without using some essential tools, called exponential identities, which are therefore, crucial. And thus, we want to know them very well. What do we mean by though, knowing them very well?

Knowing them very well, we know how they work, and more importantly, know how to work with them. So let's see now how they work and how we can work with them.

To begin with, keep in mind that all the letters used in the math expressions here take real numbers unless specified otherwise. Also, note in particular that we do not rule out cases where the letters can be 0 or negative unless specified otherwise.

Now, suppose that both a and $b \neq 0$, and that both m and n are integers.
Then, the tools below are called exponential identities, and *always* work.

0. $a^m a^n = a^{m+n}$

1. $a^m / a^n = \dfrac{a^m}{a^n} = a^{m-n}$

2. $(a^m)^n = a^{mn}$

3. $(ab)^n = a^n b^n$

4. $(a/b)^n = \left(\dfrac{a}{b}\right)^n = a^n / b^n = \dfrac{a^n}{b^n}$

Let's see now, what's going on in each of the identities listed above.

0. $a^m a^n = a^{m+n}$, where $a \neq 0$, and both m and n are integers.

To begin with, we know: multiplying 1 by a, $(m + n)$ times, we get a power a^{m+n}.

What then, do we get if multiplying a power a^m by another power a^n?

We get a new power a^{m+n}. It's because a^m is the product of m of as, and a^n is the product of n of as, so $a^m a^n$ is the product of $(m + n)$ of as.

Taking thus, the product of powers sharing the same base, we get a new power, where the base remains the same, and the exponent is the sum of the exponents. For instance:

$3^4 3^5 = (3\cdot3\cdot3\cdot3)(3\cdot3\cdot3\cdot3\cdot3) = 3\cdot3\cdot3\cdot3\cdot3\cdot3\cdot3\cdot3\cdot3 = 3^9$, so we get: $3^4 3^5 = 3^{4+5} = 3^9$.

$3^3 3^4 3^2 = (3\cdot3\cdot3)(3\cdot3\cdot3\cdot3)(3\cdot3) = 3\cdot3\cdot3\cdot3\cdot3\cdot3\cdot3\cdot3\cdot3 = 3^9$, so we get: $3^3 3^4 3^2 = 3^{3+4+2} = 3^9$.

$(-3)^2(-3)^3 = \{(-3)\cdot(-3)\}\{(-3)\cdot(-3)\cdot(-3)\} = (-3)\cdot(-3)\cdot(-3)\cdot(-3)\cdot(-3) = (-3)^5$, so we get:

$(-3)^2(-3)^3 = (-3)^{2+3} = (-3)^5$, which equals -3^5, of course.

What if m and n are negative?

Then, we have: $m < 0$, and $n < 0$, so we get: $-(m + n) > 0$. That is, $-(m + n)$ is a positive integer. So dividing 1 by a, $-(m + n)$ times, we get: $\frac{1}{a^{-(m+n)}}$.

And the denominator is a power, which is the product of $-(m + n)$ of as.

And thus, putting $\frac{1}{a^{-(m+n)}}$ in the power notation, we get: a^{m+n}, where $m + n < 0$.

So for instance, dividing 1 by a, 3 times, we get: $1/a/a/a = 1/(aaa) = \frac{1}{aaa} = \frac{1}{a^3} = a^{-3}$.

What then, do we get if multiplying a power a^m by another power a^n where m and n are negative?

We have: $-m > 0$, and $-n > 0$. So we can set: $a^m = \dfrac{1}{a^{-m}}$, and $a^n = \dfrac{1}{a^{-n}}$, where a^{-m} is the product of $-m$ of as, and a^{-n} is the product of $-n$ of as.

Thus, we get: $a^m a^n = \dfrac{1}{a^{-m}} \cdot \dfrac{1}{a^{-n}} = \dfrac{1}{a^{-(m+n)}} = a^{m+n}$.

For instance, we can get: $a^{-2} a^{-3} = \dfrac{1}{a^2} \cdot \dfrac{1}{a^3} = \dfrac{1}{a^{2+3}} = a^{-(2+3)} = a^{-5}$.

And assuming: $m = n = 0$, we get: $a^m a^n = a^0 a^0 = 1 \cdot 1 = 1$. And we have: $a^{0+0} = a^0 = 1$.

So we get: $a^m a^n = a^{m+n}$ where $a \neq 0$, and both m and n are integers. What if $a = 0$?

Then, we need to have: $m > 0$, and $n > 0$.
For instance, assuming $m = -1$, we get: $a^m = 0^{-1} = 1/0$, which is not possible.
What if $m = 0$, though?

Then, we get: $a^m = 0^0 = 1$. That's because multiplying 1 by 0, no times, we get 1.

And thus, putting threads together, we get: $a^m a^n = a^{m+n}$, where $a \neq 0$, and both m and n are integers.

1. $a^m / a^n = \dfrac{a^m}{a^n} = a^{m-n}$, where $a \neq 0$, and both m and n are integers.

We are going to look at two cases. In one, a power a^m gets divided by a, n times. And in the other, the power a^m gets divided by another power a^n.

So let's first, divide the power a^m by a, n times.
Then, assuming first, $m \geq n$, and dividing the power a^m by a, n times, we get a new power a^{m-n}, where $m - n \geq 0$. How come?

14

In the power a^m, n of as get canceled due to n divisions by a. So for instance, doing 3 divisions by a to a power a^5, we get: $\frac{a^5}{a^3} = \frac{a \cdot a \cdot a \cdot a \cdot a}{a \cdot a \cdot a} = a \cdot a = a^2 = a^{5-3}$.

- Suppose next, $m < n$.

Then, though dividing the power a^m by a, n times, we still get the power a^{m-n}, in which however, $m - n < 0$. How come?

First, if $m < n$, we can say that a^n has $(n - m)$ more as than a^m has.

For instance, since $2 < 5$, a^5 has $(5 - 2 = 3)$ more as than a^2 has.

So if $m < n$, and the power a^m gets divided by a, n times, what happens first is that all of the as in the power a^m get canceled due to the first m divisions by a, and we get 1, and then, 1 gets divided by a, $(n - m)$ more times.

Thus, we get: $\dfrac{1}{a^{n-m}}$. So we get: a^{m-n}.

So for instance, if a power a^2 gets divided by a, 5 times, what happens first is that all of the two as in the power a^2 get canceled due to the first two divisions by a, and we get 1, and then, 1 gets divided by a, $(5 - 2 = 3)$ more times.

Thus, we get: $\dfrac{1}{a^{5-2}}$. So we get: a^{5-2}.

And thus, dividing 1 by the same number, many times, we get a negative exponent, and take the same number as the base in the power we get.

So assuming: $m < n$, and dividing a^m by a, n times, we get: a^{m-n}, too. Thus, for instance, dividing a^3 by a, five times, we get: $\frac{a^3}{a^5} = \frac{a \cdot a \cdot a}{a \cdot a \cdot a \cdot a \cdot a} = \frac{1}{a \cdot a} = a^{-2} = a^{3-5}$.

So either way, dividing a^m by a, n times, we get the power, a^{m-n}.

- Suppose this time, we divide the power a^m by another power a^n. Then, we get: $\dfrac{a^m}{a^n}$.

So assuming first, $m \geq n$, we get a new power, a^{m-n}.   How come?

Initially, the numerator a^m has m of as, and the denominator a^n has n of as.

During the course of the division of a^m by a^n, n of as each in both the numerator and denominator get canceled, so the resultant numerator is left with $(m - n)$ of as, and the resultant denominator is 1.

Thus, we get: $\dfrac{a^{m-n}}{1}$, which is a^{m-n}.

So for instance, we get: $\dfrac{3^5}{3^3} = \dfrac{3\cdot3\cdot3\cdot3\cdot3}{3\cdot3\cdot3} = \dfrac{3\cdot3}{1} = 3^2 \Rightarrow \dfrac{3^5}{3^3} = 3^{5-3} = 3^2$.

And we get: $\dfrac{(\frac{1}{3})^5}{(\frac{1}{3})^3} = \dfrac{\frac{1}{3}\cdot\frac{1}{3}\cdot\frac{1}{3}\cdot\frac{1}{3}\cdot\frac{1}{3}}{\frac{1}{3}\cdot\frac{1}{3}\cdot\frac{1}{3}} = \dfrac{\frac{1}{3}\cdot\frac{1}{3}}{1} = (\frac{1}{3})^2 \Rightarrow \dfrac{(\frac{1}{3})^5}{(\frac{1}{3})^3} = (\frac{1}{3})^{5-3} = (\frac{1}{3})^2$.

And assuming next, $m < n$, we still get the power indicated by a^{m-n}.
This time though, we get: $m - n < 0$.   How come?

Initially, the numerator a^m has m of as, and the denominator a^n has n of as.

During the course of the division, m of as each in both the numerator and denominator get canceled, so the numerator ends up with 1, and the denominator is left with $(n - m)$ of as.

Thus, we get: $\dfrac{1}{a^{n-m}}$, which is the result we get if 1 gets divided by a, $(n - m)$ times.

And we know: $\dfrac{1}{a^{n-m}} = a^{-(n-m)} = a^{m-n}$.

Thus either way, dividing a power a^m by another power a^n, we get a new power a^{m-n}.

For instance, we get: $\frac{3^3}{3^5} = \frac{3 \cdot 3 \cdot 3}{3 \cdot 3 \cdot 3 \cdot 3 \cdot 3} = \frac{1}{3 \cdot 3} = \frac{1}{3^2} = 3^{-2} \Rightarrow \frac{3^3}{3^5} = 3^{3-5} = 3^{-2} = \frac{1}{3^2}$.

And we get: $\frac{\left(\frac{1}{3}\right)^3}{\left(\frac{1}{3}\right)^5} = \frac{\frac{1}{3} \cdot \frac{1}{3} \cdot \frac{1}{3}}{\frac{1}{3} \cdot \frac{1}{3} \cdot \frac{1}{3} \cdot \frac{1}{3} \cdot \frac{1}{3}} = \frac{1}{\frac{1}{3} \cdot \frac{1}{3}} = \frac{1}{\left(\frac{1}{3}\right)^2} = \left(\frac{1}{3}\right)^{-2} \Rightarrow \frac{\left(\frac{1}{3}\right)^3}{\left(\frac{1}{3}\right)^5} = \left(\frac{1}{3}\right)^{3-5} = \left(\frac{1}{3}\right)^{-2}$.

So doing a division with two powers with the same base a, we get a new power with the same base a, and the magnitude of the new exponent is the difference between the two exponents of the two powers.

And thus, we get: $a^m / a^n = \dfrac{a^m}{a^n} = a^{m-n}$, where $a \neq 0$, and both m and n are integers.

2. $(a^m)^n = a^{mn}$, where $a \neq 0$, and both m and n are integers.

Multiplying 1 by a, m times, we get a power a^m.

So multiplying 1 by the power a^m, n times, we get a new power $(a^m)^n$.

And let's see now what each multiplication can produce.

First, multiplying 1 by the power a^m, we just get: a^m, which is the product of m of as.

Next, multiplying a^m by a^m, that is, multiplying 1 by a^m, 2 times, we get: $a^m \cdot a^m = (a^m)^2$.

We know: a^m is the product of m of as.   So $a^m \cdot a^m$ is the product of $2m$ of as.

And thus, $(a^m)^2$ is the product of $2m$ of as, too.   So we get: $(a^m)^2 = a^{2m}$.

Next, multiplying a^{2m} by a^m, that is, multiplying 1 by a^m, 3 times, we get: $a^m \cdot a^m \cdot a^m = (a^m)^3$.

And we know: a^m is the product of **m** of **a**s.   So $a^m \cdot a^m \cdot a^m$ is the product of **3m** of **a**s.

And thus, $(a^m)^3$ is the product of **3m** of **a**s, too.   So we get: $(a^m)^3 = a^{3m}$.

Next, multiplying a^{2m} by a^{2m}, that is, multiplying 1 by a^m, 4 times, we get: $a^{2m} \cdot a^{2m} = (a^{2m})^2$.

We know: a^{2m} is the product of **2m** of **a**s.   So $(a^{2m} \cdot a^{2m})$ is the product of **4m** of **a**s.

And thus, $(a^{2m})^2$ is the product of **4m** of **a**s, too.   So we get: $(a^{2m})^2 = a^{4m}$.

And we have: $a^m a^n = a^m \cdot a^n = a^{m+n}$.   So we get: $a^{4m} = a^m \cdot a^m \cdot a^m \cdot a^m = (a^m)^4$.

Thus, we get: $a^{4m} = (a^m)^4$.   Now, what then, do we get multiplying 1 by a^m, **n** times?

We get: $(a^m)^n = a^{mn}$. So for instance, $(5^3)^2 = (5 \cdot 5 \cdot 5) \cdot (5 \cdot 5 \cdot 5) = 5 \cdot 5 \cdot 5 \cdot 5 \cdot 5 \cdot 5 = 5^{3 \cdot 2} = 5^6$.

And also, dividing 1 by a twice, we get: a^{-2}, which is $1/a^2$, which is $1/a/a$, too.

So dividing 1 by a^3 twice, we can get: $(a^3)^{-2} = 1/(a^3)/(a^3) = 1/(a^3 \cdot a^3) = 1/a^6 = a^{-6} = a^{3(-2)}$.

And multiplying 1 by a^{-3} twice, we can put it this way: $(a^{-3})^2$, and we get: $(a^{-3})^2 = (1/a^3)(1/a^3) = 1/(a^3 \cdot a^3) = 1/a^6 = a^{-6} = a^{3(-2)}$, too.   So we get: $(a^3)^{-2} = (a^{-3})^2 = a^{-6}$.

And dividing 1 by a^{-3} twice, we get: $(a^{-3})^{-2} = 1/(a^{-3})/(a^{-3}) = 1/(a^{-3} \cdot a^{-3}) = 1/a^{-6} = a^6$.
And we know: $a^6 = a^{(-3)(-2)}$.

   • So we get: $(a^m)^n = a^{mn}$, where $a \neq 0$, and **m** and **n** are integers.

So if a power gets raised to another power, keep the base, and take the product of the exponents.

3. $(ab)^n = a^n b^n$, where both a and $b \neq 0$, and both m and n are integers.

Multiplying 1 by a, n times, we get a power a^n.

So multiplying 1 by a product ab, n times, we get a power $(ab)^n$.

And let's see now what each multiplication can produce.

First, multiplying 1 by the product ab, we just get: ab, which is the product of 1 of as and 1 of bs.

Next, multiplying ab by ab, that is, multiplying 1 by ab, 2 times, we get:

$(ab)^2 = ab \cdot ab = aabb$, because multiplications are commutative, that is, $xy = yx$.

So we get: $(ab)^2 = a^2 b^2$.

Next, multiplying 1 by ab, 3 times, we get:

$(ab)^3 = ab \cdot ab \cdot ab = aaabbb$, since multiplications are commutative.

So we get: $(ab)^3 = a^3 b^3$.

Now, what then, do we get multiplying 1 by ab, n times?

We get: $(ab)^n = a^n b^n$. So for instance, $(3\cdot5)^3 = (3\cdot5)\cdot(3\cdot5)(3\cdot5) = 3\cdot5\cdot3\cdot5\cdot3\cdot5 = 3\cdot3\cdot3\cdot5\cdot5\cdot5 = 3^3 5^3$, and $(12\cdot7\cdot5)^2 = 12^2 7^2 5^2$.

And also, dividing 1 by ab twice, we get: $(ab)^{-2}$, which is $1/(ab)/(ab)$, which is $1/(ab)^2$, too.

So we get: $(ab)^{-2} = 1/(ab)^2 = 1/(a^2 b^2) = (1/a^2)(1/b^2) = a^{-2} b^{-2}$.

And thus, we get: $(ab)^{-2} = a^{-2} b^{-2}$.

Besides, we have: $(ab)^0 = 1$, since multiplying 1 by ab no times, we just get 1.

And we have: $a^0 = 1$, and $b^0 = 1$, so we get: $a^0 b^0 = 1$. Thus, we get: $(ab)^0 = a^0 b^0$.

So we get: $(ab)^n = a^n b^n$, where both a and $b \neq 0$, and both m and n are integers.

In short, if the base is a product, take a product of powers.

4. $(a/b)^n = \left(\dfrac{a}{b}\right)^n = a^n / b^n = \dfrac{a^n}{b^n}$, where both a and $b \neq 0$, and both m and n are integers.

Multiplying 1 by a, n times, we get a power a^n.

So multiplying 1 by a fraction a/b, n times, we get a power $(a/b)^n$.

And let's see now what each multiplication can produce.

First, multiplying 1 by the fraction a/b, we just get: a/b.

Next, multiplying a/b by a/b, that is, multiplying 1 by a/b, 2 times, we get:

$(a/b)^2 = (a/b)(a/b) = aa/(bb)$.   So we get: $(a/b)^2 = a^2/b^2$.

Next, multiplying 1 by a/b, 3 times, we get:

$(a/b)^3 = (a/b)(a/b)(a/b) = aaa/(bbb)$.   So we get: $(a/b)^3 = a^3/b^3$.

Now, what then, do we get multiplying 1 by a/b, n times?

We get: $(a/b)^n = a^n/b^n$.   So for instance, $(3/5)^3 = (3/5)(3/5)(3/5) = 3 \cdot 3 \cdot 3/(5 \cdot 5 \cdot 5) = 3^3/5^3$, and $(12/7/5)^2 = 12^2/7^2/5^2$.

And also, dividing 1 by a/b twice, we get: $(a/b)^{-2}$, which is $1/(a/b)/(a/b)$, which is $1/(a/b)^2$, too.

So we get: $(a/b)^{-2} = 1/(a/b)^2 = 1/(a^2/b^2)$.   That is, we get: $\left(\frac{a}{b}\right)^{-2} = \frac{1}{(\frac{a}{b})^2} = \frac{1}{(\frac{a^2}{b^2})}$.

And multiplying by b^2, the numerator and the denominator above, that is, multiplying the numerator 1 by b^2, and multiplying the denominator (a^2/b^2) by b^2, too, we get: b^2/a^2.

So we get: $(a/b)^{-2} = 1/(a^2/b^2) = b^2/a^2$.   That is, we get: $(a/b)^{-2} = b^2/a^2$.

And we have: $b^2 = 1/b^{-2}$, because $1/b^{-2} = b^{-(-2)} = b^2$.

And also, we have: $1/a^2 = a^{-2}$.

So we get: $b^2/a^2 = b^2(1/a^2) = (1/b^{-2})a^{-2} = a^{-2}/b^{-2}$.   Thus, we get: $b^2/a^2 = a^{-2}/b^{-2}$.

And we have: $(a/b)^{-2} = b^2/a^2$, too.   And thus, we get: $(a/b)^{-2} = a^{-2}/b^{-2}$.

Besides, we have: $(a/b)^0 = 1$, since multiplying 1 by a/b no times, we just get 1.

And we have: $a^0 = 1$, and $b^0 = 1$, so we get: $a^0/b^0 = 1$. Thus, we get: $(a/b)^0 = a^0/b^0$.

So in sum, we get: $(a/b)^n = a^n/b^n$, where both a and $b \neq 0$, and both m and n are integers.

For instance, we can get: $\left(\frac{3}{5}\right)^3 = \left(\frac{3}{5}\right)\left(\frac{3}{5}\right)\left(\frac{3}{5}\right) = \frac{3 \cdot 3 \cdot 3}{5 \cdot 5 \cdot 5} = \frac{3^3}{5^3}$.

And we can get: $\left(\frac{3}{5}\right)^{-3} = \dfrac{1}{\left(\frac{3}{5}\right)\left(\frac{3}{5}\right)\left(\frac{3}{5}\right)} = \dfrac{1}{\frac{3 \cdot 3 \cdot 3}{5 \cdot 5 \cdot 5}} = \dfrac{1}{\frac{3^3}{5^3}} = \frac{5^3}{3^3} = 5^3 \cdot \frac{1}{3^3} = \frac{1}{5^{-3}} \cdot 3^{-3} = \frac{3^{-3}}{5^{-3}}$.

Too easy? So boring? Not quite?

Right after the next section, we will get the extended versions of the identities **2**, **3**, and **4** listed above. And thus, we will get some more identities, which can give us more power when we do exponential algebra.

They are basically the same though, as the ones listed above.
The exponents *m* and *n* used in the identities above are integers, but the exponents in the extended versions are rational numbers. So the actual extensions are done on exponents.

(In fact, the exponents can be irrational, too, and thus, can be real. It is not the case though, the exponents can be any real numbers for any bases. And we will see how it is not the case in the section, **Problem Exponents**.)

Prior to the extended version, that is, in the next section, we will get familiar with another important tool, called an *n*th root, which is a solution to an equation of degree *n*, and is in fact, another form of a power.
And we are going to use the idea of such a root constructing the version extended.

Examples in Exponential Identities 1

Note that we often use a dot · instead of 'x' as the multiplication operator.

0. Convert each expression below into a simple power as 2^3.

$2 \cdot 2$, $3 \cdot 3^2$, $2 \cdot 4$, $9 \cdot 27$, $5^2 5^4$,

$4^3 4^5$, $4^5 4^{-2}$, $(-2)^3 (-2)^5$, $2^2 \div 2^3$, $2^3 \div 5^3$,

$4^3 / 4^5$, $4^5 / 4^{-2}$, $(-2)^3 / (-2)^5$, $2^{-3} / 2^3$, $2^3 / 5^{-3}$,

$(2^2)^3$, $2^3 3^3$, $2^3 / 3^3$, $2^3 / (-3)^3$, $2^3 / (-3^3)$,

$(-5^3)^2$, $2^3 3^{-3}$, $-2^3 / (-3^3)$, $-2^2 / (-3)^2$, $-2^{-3} / (-3^3)$.

1. Expand the powers below:

$(-1)^{-3}$, $(-1)^{-4}$, -1^{-4}, $(-2)^{-3}$, -2^{-3},

$(-2)^{-4}$, -0.2^{-2}, $(-0.1)^{-3}$, $(-0.1)^{-4}$, $(-0.02)^{-3}$,

2. Expand the products of powers below:

$(-0.2)^{-4}(-0.2^2)$, $-2^{-4} 3^2 0.1^2$, $(-2)^{-3}(-5^2)$, $(-5)^{-2}(-3)^2$, $(-2)^{-3}(-3^2)$.

Suggestions or Solutions
To the Examples in Exponential Identities 1

0.

$2 \cdot 2 = 2^1 \cdot 2^1 = 2^1 2^1 = 2^{1+1} = 2^2$

$3 \cdot 3^2 = 3^1 \cdot 3^2 = 3^1 3^2 = 3^{1+2} = 3^3$

$2 \cdot 4 = 2^1 \cdot 2^2 = 2^1 2^2 = 2^{1+2} = 2^3$

$9 \cdot 27 = 3^2 3^3 = 3^{2+3} = 3^5$

$5^2 5^4 = 5^{2+4} = 5^6$

$4^3 4^5 = 4^{3+5} = 4^8 = (2^2)^8 = 2^{2 \times 8} = 2^{2 \cdot 8} = 2^{16}$, and also, $4^3 4^5 = (2^2)^3 (2^2)^5 = 2^6 2^{10} = 2^{16}$

$4^5 4^{-2} = 4^{5+(-2)} = 4^{5-2} = 4^3 = (2^2)^3 = 2^{2 \cdot 3} = 2^6$, and also, $4^5 4^{-2} = (2^2)^5 (2^2)^{-2} = 2^{10} 2^{-4} = 2^6$

$(-2)^3 (-2)^5 = (-2)^{3+5} = (-2)^8 = 2^8$.
   Note that: $\mathbf{(-2)^3 = -2^3 = -8}$, but $\mathbf{(-2)^2 \neq -2^2 = -4}$, because $\mathbf{(-2)^2 = 2^2 = 4}$.

$2^2 \div 2^3 = 2^{2-3} = 2^{-1}$, and we can put it this way, too: $2^2 \div 2^3 = 2^2/2^3 = 2^{2-3} = 2^{-1}$.

$2^3 \div 5^3 = 2^3/5^3 = (2/5)^3$

$4^3/4^5 = 4^{3-5} = 4^{-2} = (2^2)^{-2} = 2^{2 \cdot (-2)} = 2^{-4}$

$4^5/4^{-2} = 4^{5-(-2)} = 4^{5+2} = 4^7 = (2^2)^7 = 2^{2 \cdot 7} = 2^{14}$

$(-2)^3/(-2)^5 = (-2)^{3-5} = (-2)^{-2} = 1/(-2)^2 = 1/2^2 = 2^{-2}$

$2^{-3}/2^3 = 2^{-3-3} = 2^{-6}$

$2^3/5^{-3} = 2^3(1/5^{-3}) = 2^3 5^3 = (2 \cdot 5)^3 = 10^3$

$(2^2)^3 = 2^{2 \cdot 3} = 2^6$

$2^3 3^3 = (2 \cdot 3)^3 = 6^3$

$2^3/3^3 = (2/3)^3$

$2^3/(-3)^3 = \{2/(-3)\}^3 = (-2/3)^3 = -(2/3)^3$, and also, $-(2/3)^3 = (-2/3)^3 = (2/-3)^3$.

$2^3/(-3^3) = 2^3/(-3)^3 = \{2/(-3)\}^3 = (-2/3)^3$, and also, $2^3/(-3^3) = (-2^3)/3^3 = -2^3/3^3 = -(2^3/3^3)$.

$(-5^3)^2 = \{(-1)5^3\}^2 = (-1)^2(5^3)^2 = (5^3)^2 = 5^6$.   So we get: $(-5^3)^2 = (5^3)^2$.

$2^3 3^{-3} = 2^3/3^3 = (2/3)^3$

$-2^3/(-3^3) = 2^3/3^3 = (2/3)^3$

$-2^2/(-3)^2 = -2^2/3^2 = -(2/3)^2$

$-2^{-3}/(-3^3) = 2^{-3}/3^3 = 1/(2^3 3^3) = 1/(2 \cdot 3)^3 = 1/6^3 = 6^{-3}$

1.

$(-1)^{-3} = 1/(-1)^3 = 1/(-1) = -1$.

$(-1)^{-4} = 1/(-1)^4 = 1/1 = 1$, and we can put it this way, too: $(-1)^{-4} = ((-1)^4)^{-1} = 1^{-1} = 1/1 = 1$.

$-1^{-4} = 1/-1^4 = 1/1 = 1$.   So we get: $-1^{-4} = (-1)^{-4} = 1$.

$(-2)^{-3} = (-2)^{3 \cdot (-1)} = ((-2)^3)^{-1} = -8^{-1} = -1/8$

$-2^{-3} = -1/2^3 = -1/8$, and of course, we can put it this way, too: $-2^{-3} = 1/-2^3 = 1/-8 = -1/8$.

$(-2)^{-4} = (-2)^{4 \cdot (-1)} = ((-2)^4)^{-1} = 16^{-1} = 1/16$, and also, we can get: $(-2)^{-4} = 1/(-2)^4 = 1/2^4$.

$-0.2^{-2} = -1/0.2^2 = 1/0.04 = 100/4 = 25$

$(-0.1)^{-3} = -1/0.1^3 = -1/0.001 = -1000$

$(-0.1)^{-4} = 1/(-0.1)^4 = 1/0.1^4 = 1/0.0001 = 10000$

$(-0.02)^{-3} = 1/(-0.02)^3 = -1/0.02^3 = -1/(0.01 \cdot 2)^3 = -1/(0.01^3 \cdot 2^3) = -1/(0.1^6 \cdot 8)$

$= -1/\{(10^{-1})^6 \cdot 8\} = -1/(10^{-6} \cdot 8) = -10^6/8 = -10^5 \cdot (10/8) = -10^5 \cdot (5/4) = -10^4 \cdot 5(10/4)$

$= -10^4 \cdot 5(5/2) = -10^4 \cdot 5^2/2 = -10^3 \cdot 5^2(10/2) = -10^3 \cdot 5^2(5) = -10^3 \cdot 5^3 = -1000 \cdot 125$

$= -125000$.

By the way, $1000/8 = 10^3/2^3 = (10/2)^3 = 5^3$.
So we get: $-10^6/8 = -10^3 10^3/2^3 = -10^3(10/2)^3 = -10^3 5^3 = -125000$, since $5^3 = 125$.
And we can put it this way, too: $-10^3 5^3 = -(10 \cdot 5)^3 = -50^3 = -125000$.

2.

$(-0.2)^{-4}(-0.2^2) = (0.2)^{-4}(-0.2^2) = -(0.2)^{-4}(0.2^2) = -0.2^{-4+2} = -0.2^{-2} = -1 \cdot 0.2^{-2} = -1/0.2^2$

$= -1/0.04 = -100/4 = -25$

$-2^{-4}3^2(0.1^2) = -1 \cdot 2^{-4}3^2(0.1^2) = -3^2(0.1^2)/2^4 = -9 \cdot 0.01/8 = -0.09/8 = -9/800$

$(-2)^{-3}(-5^2) = (2)^{-3}(5^2) = 5^2/2^3 = 25/8$

$(-5)^{-2}(-3)^2 = (5)^{-2}(3)^2 = 3^2/5^2 = 9/25$

$(-2)^{-3}(-3^2) = 2^{-3}3^2 = 3^2/2^3 = 9/8$

₂.What is an Nth Root?

Multiplying 1 by a number a particular number of times, we get a power made of a base and an exponent. For instance, multiplying 1 by 5, seven times, we get 5^7, where 5 is the base, and 7 is the exponent. In general, multiplying 1 by b, n times, we get b^n, where b is the base, and n is the exponent. And the same idea applies to divisions, too.

Dividing thus, 1 by a number a particular number of times, we get a power, too. For instance, dividing 1 by 3, eight times, we get 3^{-8}, where 3 is the base, and -8 is the exponent. In general, dividing 1 by b, n times, we get b^{-n}, where b is the base, and $-n$ is the exponent. So we get a power, too, doing divisions. That is because a division can be taken for a multiplication by the reciprocal.

What if however, we want to express a value, which has the converse nature of a power? In other words, assuming we get a value if multiplying 1 by a certain value a particular number of times, we need to find the certain value.

  • That is, given the value of a power and an exponent, we have to find the base.

Suppose for instance, multiplying 1 by a certain value seven times, we get 9.
What then, is the certain value? That is, by what value do we have to multiply 1, seven times to get 9? So we want to find a value, by which we multiply 1 seven times to get 9.

We call it a seventh root. What do we mean by though, such a root?

Saying a root in math, we can mean the solution to an equation. So finding the root, we can set up the equation describing the situation we have. And the situation is as follows:

  • Multiplying 1 by an unknown value, 7 times, we get 9.

What equation then, can we set up?

Assuming x is the unknown, we can set: $x^7 = 9$. Then, the value of x, that is, the solution to the equation above is called a 7th root.   What then, is an nth root?

It is the solution to an equation describing a situation below:

- Multiplying 1 by an unknown value, n times, we get a value called A.

So assuming again, x is the unknown, we can set up the equation the way below:

$x^n = A$ where $A > 0$, and n is a **positive integer**.

And we call the solution an nth root.   How then, can we get the solution, the nth root?

Let's first, take a look at some curves of an equation $y = x^n$ for some values of $n \geq 1$.

If $n = 1$, we just get a line passing through the origin and making $45°$ against the x-axis.

Next, if n is odd, and ≥ 3, we can put in the x-y plane, the curves the way below:

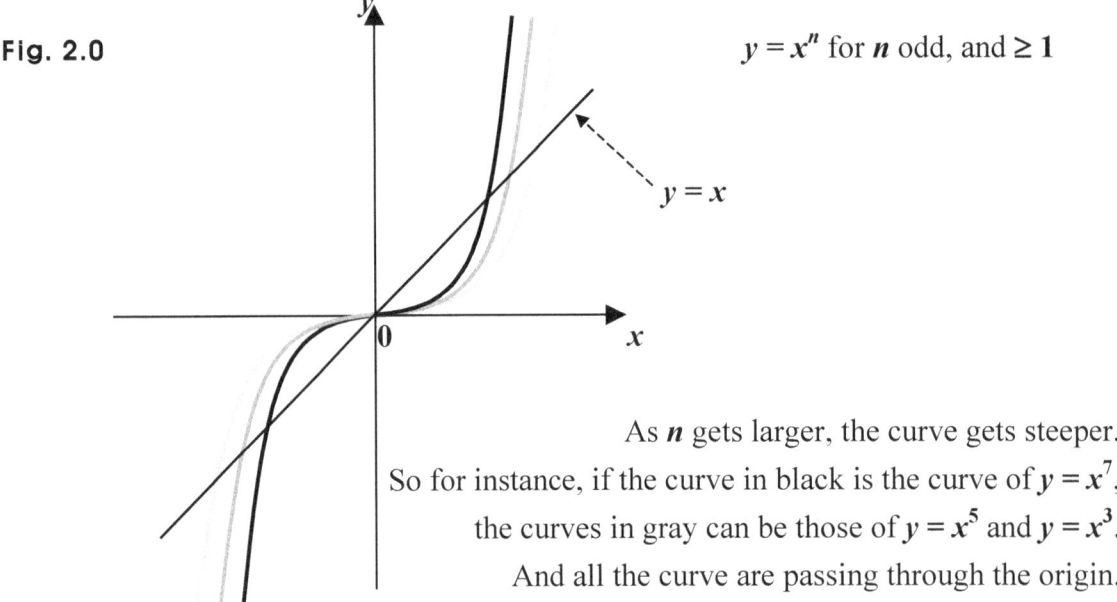

Fig. 2.0 $y = x^n$ for n odd, and ≥ 1

$y = x$

As n gets larger, the curve gets steeper.
So for instance, if the curve in black is the curve of $y = x^7$,
the curves in gray can be those of $y = x^5$ and $y = x^3$.
And all the curve are passing through the origin.

And next, if n is even, and ≥ 2, we can put in the x-y plane, the curves the way below:

Fig. 2.1

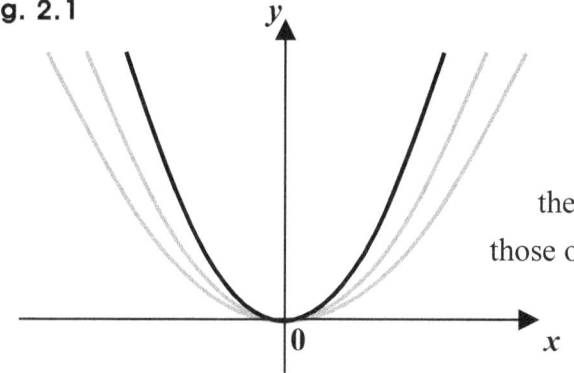

$y = x^n$ for n even, and ≥ 2

As n gets larger, the curve gets steeper. So for instance, if the curve in black is the curve of $y = x^6$, the curves in gray can be those of $y = x^4$ and $y = x^2$. And all the curves are passing through the origin.

Let's now get back to the equation: $x^n = A$ where $A > 0$, and n is a **positive integer**.

Then, considering the identity $(a^m)^n = a^{mn}$, where $a \neq 0$, and both m and n are integers, we can notice that we can put an nth root this way: $x = A^{\frac{1}{n}}$.

And putting schematically in the x-y plane, the curves of $y = x^n$, $y = A$, and $y = -A$, we can put them the way below:

Fig. 2.2

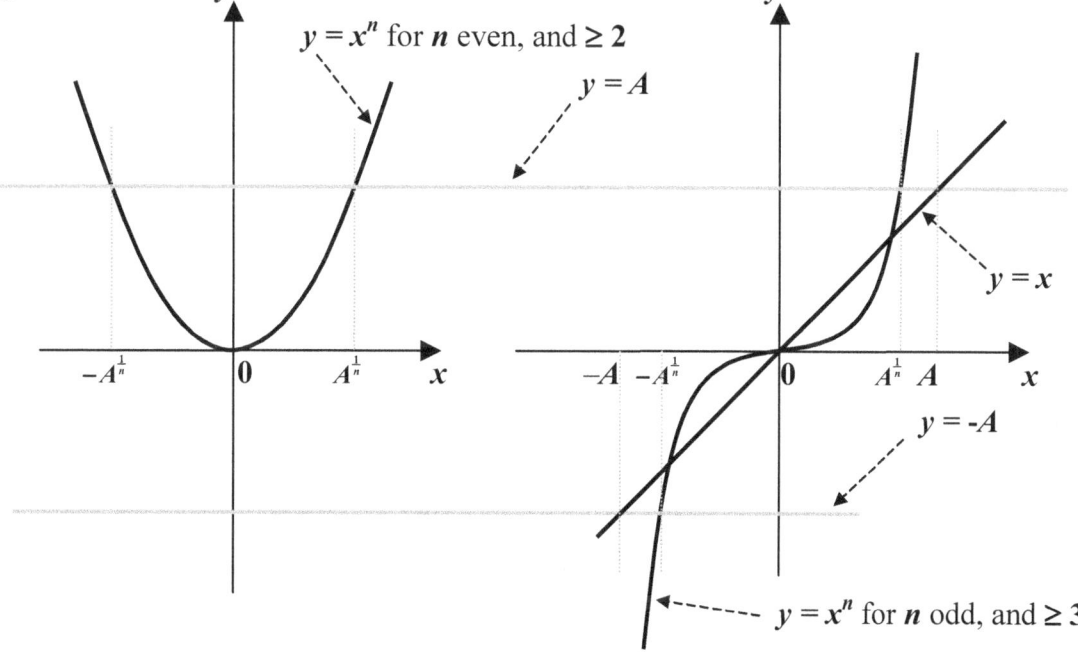

Then, we can see that the solution can be either one n^{th} root or two n^{th} roots.

It depends on the value of *n*.   So let's see now how it depends.

To begin with, if *n* = 1 or *A* = 0, there isn't much point of calling the solution an *n*th root. That's because if *n* = 1, we just get *x* = *A*, and if *A* = 0, we simply get *x* = 0.

Next, when *n* = 2 or 3, we normally call the solution a **square root** or a **cube root** rather than a second root or a third root.

What if *A* < 0, though?

Then, only if *n* is odd, we can get a solution, which is an *n*th root, too, which is negative though, as shown in the graph above. So if *n* is even, we get no solution.

- And thus, if we want to get an *n*th root for **any** integer *n* > 1, we need to have *A* > 0.

As we can see in the graph above, if *A* > 0, we can get a solution for any positive integer *n*.

So putting threads together, we can say that:

If *n* is even and *A* > 0, we get two *n*th roots, which are indicated by $A^{\frac{1}{n}}$ and $-A^{\frac{1}{n}}$.

If *n* is odd, we get one *n*th root, which is indicated by $A^{\frac{1}{n}}$.

Either way, if *A* > 0, we get the *n*th root $A^{\frac{1}{n}}$, which is positive.

- If however, *A* < 0 and *n* is even, we get no solution, that is, no root, and thus, we get no *n*th root.

So what is an *n*th root about?

It is about powers, and in fact, is a power, too.

Getting back to the equation, we have: $x^n = A$, which is saying that multiplying 1 by an unknown value, n times, we get a value called A.

And the solution is: $x = A^{\frac{1}{n}}$, which is called an nth root.

How then, can we call A and $\frac{1}{n}$?

We can call A the **base**, and can call $\frac{1}{n}$ the exponent.
So the nth root $A^{\frac{1}{n}}$ is a power, too, and thus, is about powers.

Besides, the equation $x^n = A$ is originally from the situation below:

• Multiplying 1 by a certain value, n times, we get a value called A.

And we know now that the certain value is the nth root $A^{\frac{1}{n}}$.

So multiplying 1 by the nth root $A^{\frac{1}{n}}$, n times, we get A. And multiplying 1 by the nth root $A^{\frac{1}{n}}$, n times, we get a power $(A^{\frac{1}{n}})^n$. Thus, we get: $(A^{\frac{1}{n}})^n = A$.

That is, the nth root $A^{\frac{1}{n}}$ is the certain value we want, and multiplying 1 by it, n times, we get A, which is the base in the power called the nth root, which is $A^{\frac{1}{n}}$.

Therefore, for instance, multiplying 1 by $\mathbf{9^{\frac{1}{7}}}$, 7 times, we get 9, that is, we get: $\mathbf{(9^{\frac{1}{7}})^7 = 9}$.

So we can notice now that there can be exponential identities where fractions are used as exponents. And in fact, we are going to see what such identities are, and get them, and thus, we will get to see how they work, and how to work with them in the next section.

3. Exponential Identities 2

The exponential identities covered here are the extended versions of those covered in the section, **Exponential Identities 1**, which therefore, can be called the original version in this section. So we are going to look at how we can get the original versions extended.

And getting the versions extended, we are going to use the idea of the nth root, covered in the previous section. And using the idea of the nth root, we use the structure or the configuration of such a root. What is an nth root, though?

It is a root in math, and thus, is the solution to an equation, which is of degree n.
And the equation is in fact: $x^n = A$ where $A > 0$, and n is a *positive integer* > 1.
What then, is the equation saying?

It is saying that multiplying 1 by a certain value, n times, we get A.
So solving the equation, that is, getting the solution, we get the certain value.

And we call the solution an nth root.   More specifically:

If n is even, we get two n^{th} roots, which are $A^{\frac{1}{n}}$ and $-A^{\frac{1}{n}}$.

If n is odd, we get one n^{th} root only, which is $A^{\frac{1}{n}}$.

Either way, if $A > 0$, we get the nth root $A^{\frac{1}{n}}$, which is positive.

So multiplying 1 by the nth root $A^{\frac{1}{n}}$, n times, we get A.

And also, multiplying 1 by the nth root $A^{\frac{1}{n}}$, n times, we get a power $(A^{\frac{1}{n}})^n$, which is A.

So we get: $A = (A^{\frac{1}{n}})^n$, and thus, can expect that $(A^{\frac{1}{n}})^n = A^{\frac{1}{n} \cdot n} = A$, which is in fact, the case, and is similar to an identity: $(a^m)^n = a^{mn}$, covered in **Exponential Identities 1**.

So to begin with, setting $x^n = A$, and $x > 0$, we get:
$A > 0$ and $x = A^{\frac{1}{n}}$, which is an nth root positive, and thus, we get: $(A^{\frac{1}{n}})^n = A$.

Next, getting back to the list of the original versions, and thus, assuming that both a and b are not 0, and that m and n both are integers, we get:

0. $a^m a^n = a^{m+n}$

1. $a^m / a^n = \dfrac{a^m}{a^n} = a^{m-n}$

2. $(a^m)^n = a^{mn}$

3. $(ab)^n = a^n b^n$

4. $(a/b)^n = \left(\dfrac{a}{b}\right)^n = a^n / b^n = \dfrac{a^n}{b^n}$

And next, assuming that a and b both are positive, and that m and n both are integers positive, we can put the list of the extended versions the way below:

3.1. $a^{\frac{1}{n}} b^{\frac{1}{n}} = (ab)^{\frac{1}{n}}$.

4.1. $\dfrac{a^{\frac{1}{n}}}{b^{\frac{1}{n}}} = \left(\dfrac{a}{b}\right)^{\frac{1}{n}}$.

2.1. $(a^{\frac{1}{n}})^m = (a^m)^{\frac{1}{n}}$.

2.2. $(a^{\frac{1}{n}})^{\frac{1}{m}} = a^{\frac{1}{mn}} = (a^{\frac{1}{m}})^{\frac{1}{n}}$.

2.3. $(a^{mp})^{\frac{1}{np}} = (a^m)^{\frac{1}{n}}$, where p is a positive integer.

And let see now, how we can get the extended versions above.

3.1. $a^{\frac{1}{n}} b^{\frac{1}{n}} = (ab)^{\frac{1}{n}}$.

To begin with, we have: a and $b > 0$, so we get: $ab > 0$, and thus, we get: $a^{\frac{1}{n}} b^{\frac{1}{n}} > 0$.

Next, from the original identities 3 and 2 above, we can get:

$$(a^{\frac{1}{n}} b^{\frac{1}{n}})^n = (a^{\frac{1}{n}})^n (b^{\frac{1}{n}})^n = a^{\frac{1}{n} \cdot n} \cdot b^{\frac{1}{n} \cdot n} = ab.$$

So we get: $(a^{\frac{1}{n}} b^{\frac{1}{n}})^n = ab$. And assuming: $x = a^{\frac{1}{n}} b^{\frac{1}{n}}$, and $A = ab$, we can set: $x^n = A$.

So we get: $x = A^{\frac{1}{n}}$, which is an nth root. And thus, we can get: $a^{\frac{1}{n}} b^{\frac{1}{n}} = (ab)^{\frac{1}{n}}$.

4.1. $\dfrac{a^{\frac{1}{n}}}{b^{\frac{1}{n}}} = \left(\dfrac{a}{b}\right)^{\frac{1}{n}}$.

To begin with, we have: a and $b > 0$, so we get: $\dfrac{a}{b} > 0$. And thus, we get: $\dfrac{a^{\frac{1}{n}}}{b^{\frac{1}{n}}} > 0$.

Next, from the original identities 4 and 2 above, we can get:

$$\left(\dfrac{a^{\frac{1}{n}}}{b^{\frac{1}{n}}}\right)^n = \dfrac{(a^{\frac{1}{n}})^n}{(b^{\frac{1}{n}})^n} = \dfrac{a^{\frac{1}{n} \cdot n}}{b^{\frac{1}{n} \cdot n}} = \dfrac{a}{b}.$$

So we get: $\left(\dfrac{a^{\frac{1}{n}}}{b^{\frac{1}{n}}}\right)^n = \dfrac{a}{b}$. And assuming: $x = \dfrac{a^{\frac{1}{n}}}{b^{\frac{1}{n}}}$, and $A = \dfrac{a}{b}$, we can set: $x^n = A$.

So we get: $x = A^{\frac{1}{n}}$, which is an nth root. And thus, we can get: $\dfrac{a^{\frac{1}{n}}}{b^{\frac{1}{n}}} = \left(\dfrac{a}{b}\right)^{\frac{1}{n}}$.

2.1. $(a^{\frac{1}{n}})^m = (a^m)^{\frac{1}{n}}$.

To begin with, we have: $a > 0$, so we get: $a^{\frac{1}{n}} > 0$, and thus, we get: $(a^{\frac{1}{n}})^m > 0$.

And since we have: $a > 0$, we get: $a^m > 0$, too.

Next, from the original identity 2 above, we can get:

$$\{(a^{\frac{1}{n}})^m\}^n = (a^{\frac{1}{n}})^{mn} = (a^{\frac{1}{n}})^{nm} = \{(a^{\frac{1}{n}})^n\}^m = (a^{\frac{1}{n}\cdot n})^m = a^m.$$

Thus, we get $\{(a^{\frac{1}{n}})^m\}^n = a^m$. And assuming: $x = (a^{\frac{1}{n}})^m$, and $A = a^m$, we can set: $x^n = A$.

So we get: $x = A^{\frac{1}{n}}$, which is an nth root. And thus, we can get: $(a^{\frac{1}{n}})^m = (a^m)^{\frac{1}{n}}$.

And of course, from the original identity 2 above, we can get the same this way, too:

$$(a^m)^{\frac{1}{n}} = a^{\frac{m}{n}} = (a^{\frac{1}{n}})^m.$$

2.2. $(a^{\frac{1}{n}})^{\frac{1}{m}} = a^{\frac{1}{mn}} = (a^{\frac{1}{m}})^{\frac{1}{n}}$.

To begin with, we have: $a > 0$, so we get: $(a^{\frac{1}{n}})^{\frac{1}{m}} > 0$, and $(a^{\frac{1}{m}})^{\frac{1}{n}} > 0$.

Next, from the original identity 2 above, we can get:

$$\{(a^{\frac{1}{n}})^{\frac{1}{m}}\}^{mn} = [\{(a^{\frac{1}{n}})^{\frac{1}{m}}\}^m]^n = \{(a^{\frac{1}{n}})^{\frac{m}{m}}\}^n = (a^{\frac{1}{n}})^n = a, \text{ and also:}$$

$$\{(a^{\frac{1}{m}})^{\frac{1}{n}}\}^{mn} = \{(a^{\frac{1}{m}})^{\frac{1}{n}}\}^{nm} = [\{(a^{\frac{1}{m}})^{\frac{1}{n}}\}^n]^m = \{(a^{\frac{1}{m}})^{\frac{n}{n}}\}^m = (a^{\frac{1}{m}})^m = a.$$

So we get: $\{(a^{\frac{1}{n}})^{\frac{1}{m}}\}^{mn} = a$, and $\{(a^{\frac{1}{m}})^{\frac{1}{n}}\}^{mn} = a$.

Now, the product **mn** is another integer positive, since **m** and **n** both are integers positive.

And assuming: $x = (a^{\frac{1}{n}})^{\frac{1}{m}}$, and $A = a$, we can set: $x^{mn} = A$.

So we get: $x = A^{\frac{1}{mn}}$, which is an **mn**th root. Thus, we can get: $(a^{\frac{1}{n}})^{\frac{1}{m}} = a^{\frac{1}{mn}}$.

And also, assuming: $x = (a^{\frac{1}{m}})^{\frac{1}{n}}$, and $A = a$, we can set again: $x^{mn} = A$.

So we get: $x = A^{\frac{1}{mn}}$, too, which is an **mn**th root, also. Thus, we can get: $(a^{\frac{1}{m}})^{\frac{1}{n}} = a^{\frac{1}{mn}}$, too.

And thus, putting threads together, we get: $(a^{\frac{1}{n}})^{\frac{1}{m}} = a^{\frac{1}{mn}} = (a^{\frac{1}{m}})^{\frac{1}{n}}$.

2.3. $(a^{mp})^{\frac{1}{np}} = (a^m)^{\frac{1}{n}}$, where p is a positive integer.

To begin with, we have: $a > 0$, so we get: $a^m > 0$, and thus, we can get: $(a^{mp})^{\frac{1}{np}} > 0$.

Next, from the original identity 2 above, we can get:

$$\{(a^{mp})^{\frac{1}{np}}\}^n = [\{(a^{mp})^{\frac{1}{p}}\}^{\frac{1}{n}}]^n = \{(a^{mp})^{\frac{1}{p}}\}^{\frac{n}{n}} = (a^{mp})^{\frac{1}{p}} = \{(a^m)^p\}^{\frac{1}{p}} = (a^m)^{\frac{p}{p}} = a^m.$$

So we get: $\{(a^{mp})^{\frac{1}{np}}\}^n = a^m$. And assuming: $x = (a^{mp})^{\frac{1}{np}}$, and $A = a^m$, we can set: $x^n = A$.

So we get: $x = A^{\frac{1}{n}}$, which is an nth root. Thus, we can get: $(a^{mp})^{\frac{1}{np}} = (a^m)^{\frac{1}{n}}$.

What if though, **m** and **n** both are integers negative?

Suppose first, $x^k = \frac{1}{A}$, where $A > 0$, and k is an integer positive.

Then, we get: $\frac{1}{A} > 0$, so we can get: $x = \left(\frac{1}{A}\right)^{\frac{1}{k}}$, which is a kth (that is, n^{th}) root.

Next, we have: $\frac{1}{A} = A^{-1}$. So we get: $x = \left(\frac{1}{A}\right)^{\frac{1}{k}} = (A^{-1})^{\frac{1}{k}} = A^{-\frac{1}{k}} = A^{\frac{1}{-k}} \Rightarrow x = A^{\frac{1}{-k}}$.

And we have: $x = \left(\frac{1}{A}\right)^{\frac{1}{k}}$.

So we get: $A^{-\frac{1}{k}} = \left(\frac{1}{A}\right)^{\frac{1}{k}}$, which is a kth root, and thus, $A^{-\frac{1}{k}}$ is a kth root, too.

Now, k is an integer positive, so $-k$ is an integer negative.

Also, n in the word, 'n^{th} root' is an integer, too.

So we can see that the *idea* of an nth root works for $n < 0$, too, in the extended versions.

And getting back to the list of the extended versions, and thus, assuming that a and b both are positive, and that m and n both are integers positive, we get:

3.1. $a^{\frac{1}{n}}b^{\frac{1}{n}} = (ab)^{\frac{1}{n}}$. **4.1.** $\dfrac{a^{\frac{1}{n}}}{b^{\frac{1}{n}}} = \left(\dfrac{a}{b}\right)^{\frac{1}{n}}$. **2.1.** $(a^{\frac{1}{n}})^{m} = (a^{m})^{\frac{1}{n}}$.

2.2. $(a^{\frac{1}{n}})^{\frac{1}{m}} = a^{\frac{1}{mn}} = (a^{\frac{1}{m}})^{\frac{1}{n}}$. **2.3.** $(a^{mp})^{\frac{1}{np}} = (a^{m})^{\frac{1}{n}}$, where p is a positive integer.

• Suppose now, we replace a with $\frac{1}{a}$, and b with $\frac{1}{b}$ in the extended versions.

Then, we'll get to see $\frac{1}{-m}$, $\frac{1}{-n}$, and $-m$ instead of $\frac{1}{m}$, $\frac{1}{n}$, and m.

For instance, in $a^{\frac{1}{n}}b^{\frac{1}{n}} = (ab)^{\frac{1}{n}}$, replacing a and b with $\frac{1}{a}$ and $\frac{1}{b}$ respectively, we get:

$a^{\frac{1}{n}}b^{\frac{1}{n}} = (ab)^{\frac{1}{n}} \Rightarrow (\frac{1}{a})^{\frac{1}{n}}(\frac{1}{b})^{\frac{1}{n}} = (\frac{1}{a} \cdot \frac{1}{b})^{\frac{1}{n}}$.

Meanwhile:

$(\frac{1}{a})^{\frac{1}{n}}(\frac{1}{b})^{\frac{1}{n}} = (a^{-1})^{\frac{1}{n}}(b^{-1})^{\frac{1}{n}} = a^{-\frac{1}{n}}b^{-\frac{1}{n}} = a^{\frac{1}{-n}}b^{\frac{1}{-n}}$.

$(\frac{1}{a} \cdot \frac{1}{b})^{\frac{1}{n}} = (a^{-1}b^{-1})^{\frac{1}{n}} = \{(ab)^{-1}\}^{\frac{1}{n}} = (ab)^{-\frac{1}{n}} = (ab)^{\frac{1}{-n}}$.

So we can get: $a^{\frac{1}{n}}b^{\frac{1}{n}} = (ab)^{\frac{1}{n}} \Rightarrow a^{\frac{1}{-n}}b^{\frac{1}{-n}} = (ab)^{\frac{1}{-n}}$.

And by the same token, doing the same replacement as above, we will get to see that:

4.1. $\dfrac{a^{\frac{1}{n}}}{b^{\frac{1}{n}}}=\left(\dfrac{a}{b}\right)^{\frac{1}{n}} \Rightarrow \dfrac{a^{\frac{1}{-n}}}{b^{\frac{1}{-n}}}=\left(\dfrac{a}{b}\right)^{\frac{1}{-n}}.$

2.1. $(a^{\frac{1}{n}})^{m}=(a^{m})^{\frac{1}{n}} \Rightarrow (a^{\frac{1}{-n}})^{m}=(a^{m})^{\frac{1}{-n}}.$

2.2. $(a^{\frac{1}{n}})^{\frac{1}{m}}=a^{\frac{1}{mn}}=(a^{\frac{1}{m}})^{\frac{1}{n}} \Rightarrow (a^{\frac{1}{-n}})^{\frac{1}{-m}}=a^{\frac{1}{mn}}=(a^{\frac{1}{-m}})^{\frac{1}{-n}}.$

2.3. $(a^{mp})^{\frac{1}{np}}=(a^{m})^{\frac{1}{n}} \Rightarrow (a^{-mp})^{\frac{1}{np}}=(a^{-m})^{\frac{1}{n}},$ where p is a positive integer.

Now, we have: $a > 0$, and $b > 0$. So we get: $\frac{1}{a} > 0$, and $\frac{1}{b} > 0$.

And thus, setting $c > 0$ and $d > 0$, and replacing $\frac{1}{a}$ and $\frac{1}{b}$ with c and d, we can make another set of versions separate from the extended versions.

We don't have to bother doing that, though. The extended versions use a and b, which cover all numbers positive. So we can just use the extended versions for that purpose, too.   What then, about m and n?

We know that in $\frac{1}{-m}$ and $\frac{1}{-n}$, $-m$ and $-n$ are negative integers.

So we just let m and n in the extended versions take integers negative, too.
In other words, we can just replace 'positive' with 'nonzero' in the extended versions.
So we can put the extended versions the way below:

Assuming that a and $b > 0$, and that m, n, and p are *integers nonzero*, we get:

3.1. $a^{\frac{1}{n}}b^{\frac{1}{n}}=(ab)^{\frac{1}{n}}.$ **4.1.** $\dfrac{a^{\frac{1}{n}}}{b^{\frac{1}{n}}}=\left(\dfrac{a}{b}\right)^{\frac{1}{n}}.$ **2.1.** $(a^{\frac{1}{n}})^{m}=(a^{m})^{\frac{1}{n}}.$

2.2. $(a^{\frac{1}{n}})^{\frac{1}{m}}=a^{\frac{1}{mn}}=(a^{\frac{1}{m}})^{\frac{1}{n}}.$ **2.3.** $(a^{mp})^{\frac{1}{np}}=(a^{m})^{\frac{1}{n}}.$

What if $a < 0$ or $b < 0$, though?

Then, the extended versions might not work.

In fact, if in a power, the base is negative and the exponent is a fraction where the denominator is even, the power is very likely to be in trouble. How come?

The same issue will be raised in the next section, too, and the detailed explanations will be presented in the sections for **Problem Bases**.

Examples in Exponential Identities 2

Note that as the multiplication operator, we often use a dot · instead of 'x' as in 2·7 = 14.

0.   Put each expression below in a simple power as 2^3.

$2^2 \cdot 2^{0.5}$, $2^{0.2} \cdot 2^{-0.1}$, $2^{10.2} \cdot 2^{0.8}$, $2^2 \div 2^{1/2}$, $2^{0.2}/2^{-0.1}$,

$2^{0.2}/2^{-0.1}$, $2^{10.2}/2^{0.8}$, $2^{10.2}/2^{0.8}$, $(2^2)^{1/2}$, $(2^{0.2})^{-0.1}$,

$(2^{10.2})^{0.8}$, $2^{1/2}3^{1/2}$, $2^{-0.1}3^{-0.1}$, $2^{0.8}3^{0.8}$, $2^{1/2}/3^{1/2}$,

$2^{-0.1}/3^{-0.1}$, $2^{0.8}/2^{0.8}$, $(16^{\frac{1}{3}})^{\frac{1}{4}}(16^{\frac{1}{3}})^{\frac{1}{2}}$, $(\boldsymbol{a^2})^{1/5}\boldsymbol{a}^{1/3}$.

1.   Simplify the expressions below.

For example, $\boldsymbol{a^3}$ x $\boldsymbol{3a^2b}$ x $\boldsymbol{2b^3}$ = $\boldsymbol{a^3 \cdot 3a^2b \cdot 2b^3}$ = $\boldsymbol{a^3 3a^2 b 2b^3}$ = $\boldsymbol{6a^5 b^4}$.

1.0 $\boldsymbol{a^3}$ x $\boldsymbol{3a^2 b}$

1.1 $\boldsymbol{3^2 ab^2}$ x $\boldsymbol{2a^2 b^3}$

1.2 $\boldsymbol{2a^3 b^2 c}$ x $\boldsymbol{3^2 a^7 b^3 c^2}$ x $\boldsymbol{5a^2 b^5 c^2 d^2}$

1.3 $\boldsymbol{2b^2 c^4 e^3}$ x $\boldsymbol{3^2 a^2 b^3 c^2}$ x $\boldsymbol{2^7 c^2 d^5 e^2}$ x $\boldsymbol{2^5 a^3 b^7 c^3 d}$ x $\boldsymbol{3^5 a^5 d^3 e^4}$

1.4 $\boldsymbol{(a^3 b^2 c)^3}$ x $\boldsymbol{bc^3 d}$

1.5 $a^3bd^3 \times (a^3b)^3$

1.6 $(3a^2b^4d^3)^2 \times (16^3ac^3d)^5$

1.7 $15^3(a^3b^2c)^4 \times 26^3(4^5a^2b)^5c^3d$

1.8 $(45^3ab)^7c^3d \times 625(a^3b)^2d^3 \times (234^3a^5b)^3e$

1.9 $3b^2ea^3 \times (-12^2d^2b)^2 \times 9^2ad^2 \times (-4b^2d^4e^2c)^4 \times (9^3a^2bc^2)^8$

1.A $(32^2b^2ad^2)^7 \times (-425^2c^2ae^2d)^3 \times (-81b^2ca^2)^5 \times (-244^2ea^2)^{12} \times (49^2be^2a^3)^3$

1.B $(3b^2ea^3)^{-3} \times (2^2d^{-2}e^2b)^3 \div (-6^2ac^2d^2)^7 \times (4b^2d^{-3}e^2c)^{-2} \times (-9^2a^2bc^2)^{-9}$

Suggestions or Solutions
To the Examples in Exponential Identities 2

0.

$2^2 \cdot 2^{0.5} = 2^{2+0.5} = 2^{2.5}$

$2^{0.2} \cdot 2^{-0.1} = 2^{0.2 + (-0.1)} = 2^{0.1}$

$2^{10.2} \cdot 2^{0.8} = 2^{10.2 + 0.8} = 2^{11}$

$2^2 \div 2^{1/2} = 2^{2 - 1/2} = 2^{3/2}$, or we can put it this way, too: $2^2 \div 2^{1/2} = 1/2^{1/2 - 2} = 1/2^{-3/2} = 2^{3/2}$

$2^{0.2}/2^{-0.1} = 2^{0.2 - (-0.1)} = 2^{0.3}$

$2^{0.2}/2^{-0.1} = 1/2^{(-0.1) - 0.2} = 1/2^{-0.3} = 2^{0.3}$

$2^{10.2} / 2^{0.8} = 2^{10.2 - 0.8} = 2^{9.4}$

$2^{10.2} / 2^{0.8} = 1/2^{0.8 - 10.2} = 1/2^{-9.4} = 2^{9.4}$

$(2^2)^{1/2} = 2^{2 \times 1/2} = 2^1 = 2$

$(2^{0.2})^{-0.1} = 2^{0.2 \times (-0.1)} = 2^{-0.02}$

$(2^{10.2})^{0.8} = 2^{10.2 \times 0.8} = 2^{8.16}$

$2^{1/2}3^{1/2} = 2^{1/2} \cdot 3^{1/2} = (2 \cdot 3)^{1/2} = 6^{1/2}$

$2^{-0.1}3^{-0.1} = 2^{-0.1} \times 3^{-0.1} = (2 \cdot 3)^{-0.1} = 6^{-0.1}$

$2^{0.8}3^{0.8} = 2^{0.8} \cdot 3^{0.8} = (2 \cdot 3)^{0.8} = 6^{0.8}$

$2^{1/2}/3^{1/2} = (2/3)^{1/2}$

$2^{-0.1}/3^{-0.1} = (2/3)^{-0.1}$

$(2/3)^{0.8} = 2^{0.8}/2^{0.8}$

$(16^{\frac{1}{3}})^{\frac{1}{4}}(16^{\frac{1}{3}})^{\frac{1}{2}} = (16^{1/4})^{1/3}(16^{1/2})^{1/3} = 2^{1/3}4^{1/3} = 8^{1/3} = 2.$

$(a^2)^{1/5}a^{1/3} = (a^2)^{3/15}a^{5/15} = (a^6)^{1/15}(a^5)^{1/15} = (a^6a^5)^{1/15} = (a^{6+5})^{1/15} = (a^{11})^{1/15} = a^{11/15}.$

And we can put it this way, too: $(a^2)^{1/5}a^{1/3} = a^{2/5}a^{1/3} = a^{6/15}a^{5/15} = a^{(6+5)/15} = a^{11/15}.$

1.

1.0 $a^3 \times 3a^2b = 3a^{3+2}b = 3a^5b$

1.1 $3^2ab^2 \times 2a^2b^3 = 3^2 \times 2a^{1+2} \times b^{2+3} = 2 \times 3^2a^3b^5 = 2\cdot3^2a^3b^5$

1.2 $2a^3b^2c \times 3^2a^7b^3c^2 \times 5a^2b^5c^2d^2 = 2 \times 3^2 \times 5a^{3+7+2}b^{2+3+5}c^{1+2+2}d^2 = 2\cdot3^2\cdot5a^{12}b^{10}c^5d^2$

1.3 $2b^2c^4e^3 \times 3^2a^2b^3c^2 \times 2^7c^2d^5e^2 \times 2^5a^3b^7c^3d \times 3^5a^5d^3e^4$

$= 2^{1+7+5}3^{2+5}a^{2+3+5}b^{2+3+7}c^{4+2+2+3}d^{5+1+3}e^{3+2+4} = 2^{13}3^7a^{10}b^{12}c^{11}d^9e^9$

1.4 $(a^3b^2c)^3 \times bc^3d$

To begin with, we can get: $(a^3b^2c)^3 = a^{3\times3}b^{2\times3}c^{1\times3} = a^9b^6c^3$

So, we get: $(a^3b^2c)^3 \times bc^3d = a^9b^6c^3 \times bc^3d = a^9b^{6+1}c^{3+3}d = a^9b^7c^6d.$

1.5 $a^3bd^3 \times (a^3b)^3$

To begin with, we can get: $(a^3b)^3 = a^{3\times3}b^{1\times3} = a^9b^3$.

So we get: $a^3bd^3 \times (a^3b)^3 = a^3bd^3 \times a^9b^3 = a^{3+9}b^{1+3}d^3 = a^{12}b^4d^3$

1.6 $(3a^2b^4d^3)^2 \times (16^3ac^3d)^5 = 3^{1\times2}a^{2\times2}b^{4\times2}d^{3\times2} \times 16^{3\times5}a^{1\times5}c^{3\times5}d^5$

$= 3^2a^4b^8d^6 \times (2^4)^{15}a^5c^{15}d^5 = 3^2a^4b^8d^6 \times 2^{4\times15}a^5c^{15}d^5 = 2^{60}3^2a^{4+5}b^8c^{15}d^{6+5} = 2^{60}3^2a^9b^8c^{15}d^{11}$

1.7 $15^3(a^3b^2c)^4 \times 26^3(4^5a^2b)^5c^3d = (3 \times 5)^3a^{3\times4}b^{2\times4}c^{1\times4} \times (2 \times 13)^34^{5\times5}a^{2\times5}b^{1\times5}c^3d$

$= 3^55^3a^{12}b^8c^4 \times 2^313^34^{25}a^{10}b^5c^3d = 2^33^55^313^3a^{12+10}b^{8+5}c^{4+3}d(2^2)^{25}$

$= 2^{3+2\times25}3^55^313^3a^{22}b^{13}c^7d = 2^{53}3^55^313^3a^{22}b^{13}c^7d$

1.8 $(45^3ab)^7c^3d \times 625(a^3b)^2d^3 \times (234^3a^5b)^3e$

To begin with, we can get:

$45 = 5 \times 9 = 5 \times 3^2$, $45^3 = 5^{1\times3}3^{2\times3} = 5^33^6$, $625 = 5^4$,

$234 = 2 \times 117 = 2 \times 3 \times 39 = 2 \times 3 \times 3 \times 13 = 2 \times 3^2 \times 13$, and

$234^3 = (2 \times 3^2 \times 13)^3 = 2^{1\times2}3^{2\times3}13^{1\times3} = 2^23^613^3$

So we get: $(45^3ab)^7c^3d$ x $625(a^3b)^2d^3$ x $(234^3a^5b)^3e$

$= (5^33^6)^7a^{1\times7}b^{1\times7}c^3d$ x $5^4a^{3\times2}b^{1\times2}d^3$ x $(2^23^613^3)^3a^{5\times3}b^{1\times3}e$

$= 5^{3\times7}3^{6\times7}a^7b^7c^3d$ x $5^4a^6b^2d^3$ x $2^{2\times3}3^{6\times3}13^9a^{15}b^3e$

$= 2^63^{42+18}5^{21+4}13^9a^{7+6+15}b^{7+2+3}c^3d^{1+3}e$

$= 2^63^{60}5^{25}13^9a^{28}b^{12}c^3d^4e$

1.9 $3b^2ea^3$ x $(-12^2d^2b)^2$ x 9^2ad^2 x $(-4b^2d^4e^2c)^4$ x $(9^3a^2bc^2)^8$

First, we have: $(-1)^2 = 1$.
In fact, $(-1)^n = 1$ if n is an even integer, and if n is odd, $(-1)^n = 1$.

Next, $-12^2d^2b = (-1)$ x 12^2d^2b, and $12 = 3$ x $4 = 2^2$ x $3 = 2^23$.

So next, $(-12^2d^2b)^2 = 12^{2\times2}d^{2\times2}b^{1\times2} = (2^23)^4d^4b^2 = 2^{2\times4}3^{1\times4}d^4b^2$, and $9^2ad^2 = 3^4ad^2$.

Next, $(-4b^2d^4e^2c)^4 = 2^{2\times4}d^{4\times4}e^{2\times4}c^{1\times4} = 2^8d^{16}e^8c^4$.

And next, $(9^3a^2bc^2)^8 = (3^6a^2bc^2)^8 = 3^{6\times8}a^{2\times8}b^{1\times8}c^{2\times8}$.

So we get: $3b^2ea^3$ x $(-12^2d^2b)^2$ x 9^2ad^2 x $(-4b^2d^4e^2c)^4$ x $(9^3a^2bc^2)^8$

$= 3b^2ea^3$ x $2^83^4d^4b^2$ x 3^4ad^2 x $2^8d^{16}e^8c^4$ x $3^{48}a^{16}b^8c^{16}$

$= 2^{8+8}3^{1+4+4+48}a^{3+1+16}b^{2+2+8}c^{4+16}d^{4+2+16}e^{1+8}$

$= 2^{16}3^{57}a^{20}b^{12}c^{20}d^{22}e^9$.

1.A $(32^2b^2ad^2)^7$ x $(-425^2c^2ae^2d)^3$ x $(-81b^2ca^2)^5$ x $(-244^2ea^2)^{12}$ x $(49^2be^2a^3)^3$

To begin with, $32^2 = 2^{5x2} = 2^{10}$.

Next, $(32^2b^2ad^2)^7 = 2^{10x7}b^{2x7}a^{1x7}d^{2x7} = 2^{70}b^{14}a^7d^{14}$.

Next, $425 = (5 \times 85) = (5 \times 5 \times 17) = 5^2 \times 17 = 5^217$.

So we get: $425^2 = (5^2 \times 17)^2 = 5^{2x2} \times 17^{1x2} = 5^4 \times 17^2 = 5^417^2$.

And we have: $(-1)^3 = -1$.

So we get: $(-425^2c^2ae^2d)^3 = -5^{4x3}17^{2x3}c^{2x3}a^{1x3}e^{2x3}d^{1x3} = -5^{12}17^6c^6a^3e^6d^3$.

Next, $(-1)^5 = -1$ and $81 = 3^4$, so we get: $(-81b^2ca^2)^5 = -3^{4x5}b^{2x5}c^{1x5}a^{2x5} = -3^{20}b^{10}c^5a^5$.

Next, $(-1)^{12} = 1$, $244 = 2 \times 122 = 2 \times 2 \times 61 = 2^2 \times 61$, and $244^2 = 2^{2x2} \times 61^{1x2} = 2^461^2$.

So we get: $(-244^2ea^2)^{12} = 2^{4x12}61^{2x12}e^{1x12}a^{2x12} = 2^{48}61^{24}e^{12}a^{24}$.

And next, $49 = 7^2$ and $49^2 = 7^4$, so we get: $(49^2be^2a^3)^3 = 7^{4x3}b^{1x3}e^{2x3}a^{3x3} = 7^{12}b^3e^6a^9$.

Thus, we get:

$(32^2b^2ad^2)^7$ x $(-425^2c^2ae^2d)^3$ x $(-81b^2ca^2)^5$ x $(-244^2ea^2)^{12}$ x $(49^2be^2a^3)^3$

$= 2^{70}b^{14}a^7d^{14}$ x $(-5^{12}17^6c^6a^3e^6d^3)$ x $(-3^{20}b^{10}c^5a^5)$ x $2^{48}61^{24}e^{12}a^{24}$ x $7^{12}b^3e^6a^9$

$= 2^{70+48}3^{20}5^{12}7^{12}17^661^{24}a^{7+3+5+24+9}b^{14+10+3}c^{6+5}d^{14+3}e^{6+12+6}$

$= 2^{118}3^{20}5^{12}7^{12}17^661^{24}a^{48}b^{27}c^{11}d^{17}e^{24}$.

1.B $(3b^2ea^3)^{-3}$ x $(2^2d^{-2}e^2b)^3$ ÷ $(-6^2ac^2d^2)^7$ x $(4b^2d^{-3}e^2c)^{-2}$ x $(-9^2a^2bc^2)^{-9}$

To begin with, we can get: $(3b^2ea^3)^{-3} = 3^{1\times(-3)}b^{2\times(-3)}e^{1\times(-3)}a^{3\times(-3)} = 3^{-3}b^{-6}e^{-3}a^{-9}$.

Next, $(2^2d^{-2}e^2b)^3 = 2^{2\times3}d^{(-2)\times3}e^{2\times3}b^{1\times3} = 2^6d^{-6}e^6b^3$.

Next, $(-1)^7 = -1$, and $6^2 = 2^2$ x 3^2.

So we get: $(-6^2ac^2d^2)^7 = -2^{2\times7}3^{2\times7}a^{1\times7}c^{2\times7}d^{2\times7} = -2^{14}3^{14}a^7c^{14}d^{14}$.

Next, $(4b^2d^{-3}e^2c)^{-2} = 2^{2\times(-2)}b^{2\times(-2)}d^{(-3)\times(-2)}e^{2\times(-2)}c^{1\times(-2)} = 2^{-4}b^{-4}d^6e^{-4}c^{-2}$.

Next, $(-1)^{-9} = -1$, so we get: $(-9^2a^2bc^2)^{-9} = -3^{4\times(-9)}a^{2\times(-9)}b^{1\times(-9)}c^{2\times(-9)} = -3^{-36}a^{-18}b^{-9}c^{-18}$.

Thus, we get: $(3b^2ea^3)^{-3}$ x $(2^2d^{-2}e^2b)^3$ ÷ $(-6^2ac^2d^2)^7$ x $(4b^2d^{-3}e^2c)^{-2}$ x $(-9^2a^2bc^2)^{-9}$

$= 3^{-3}b^{-6}e^{-3}a^{-9}$ x $2^6d^{-6}e^6b^3$ ÷ $(-2^{14}3^{14}a^7c^{14}d^{14})$ x $2^{-4}b^{-4}d^6e^{-4}c^{-2}$ x $(-3^{-36}a^{-18}b^{-9}c^{-18})$.

And next, we have: $1/A = A^{-1}$ where $A \neq 0$.

So we get: $3^{-3}b^{-6}e^{-3}a^{-9}$ x $2^6d^{-6}e^6b^3$ ÷ $(-2^{14}3^{14}a^7c^{14}d^{14})$ x $2^{-4}b^{-4}d^6e^{-4}c^{-2}$ x $(-3^{-36}a^{-18}b^{-9}c^{-18})$

$= 3^{-3}b^{-6}e^{-3}a^{-9}$ x $2^6d^{-6}e^6b^3$ x $(-2^{14}3^{14}a^7c^{14}d^{14})^{-1}$ x $2^{-4}b^{-4}d^6e^{-4}c^{-2}$ x $(-3^{-36}a^{-18}b^{-9}c^{-18})$

$= 3^{-3}b^{-6}e^{-3}a^{-9}$ x $2^6d^{-6}e^6b^3$ x $(-2^{-14}3^{-14}a^{-7}c^{-14}d^{-14})$ x $2^{-4}b^{-4}d^6e^{-4}c^{-2}$ x $(-3^{-36}a^{-18}b^{-9}c^{-18})$

$= 2^{6-14-4}3^{-3-14-36}a^{-9-7-18}b^{-6+3-4-9}c^{-14-2-18}d^{6-14+6}e^{-3+6-4}$

$= 2^{-12}3^{-53}a^{-34}b^{-16}c^{-34}d^{14}e^{-1}$.

4. What are Radical Signs?

Expressing an nth root, we can put it in power notation as in $A^{\frac{1}{n}}$. It's not the only way though. We can indicate it by means of a radical sign, too. What then, is a radical?

It is just another word for a *root* in math, and thus, can be said to be a solution to an equation as $x^2 - x - 2 = 0$. So an nth root can be called an nth radical, too. And thus, if a radical sign is used to indicate an nth root, it is called an nth root sign or an nth radical sign, and has such a simple syntax as follows:

$\sqrt[n]{}$, where n is a positive integer ≥ 2, and indicates the degree of the radical.

So we can set: $A^{\frac{1}{n}} = \sqrt[n]{A}$, and we get: $(\sqrt[n]{A})^n = A$.

And thus, we call $\sqrt[n]{A}$ an nth root of A or an nth radical of A.

For instance, we can put $9^{\frac{1}{7}}$ in $\sqrt[7]{9}$, which can be called a seventh root (radical) of 9.

So $9^{\frac{1}{7}}$ (9 to the one seventh) $= \sqrt[7]{9}$ (a seventh root of 9), and we get: $(\sqrt[7]{9})^7 = 9$.

For another instance, $\sqrt[4]{16}$ can be called a fourth root of 16, and we get: $(\sqrt[4]{16})^4 = 16$.

In addition, we have: $16 = 2^4$ and $\sqrt[4]{16} = 16^{\frac{1}{4}}$, so we get: $\sqrt[4]{16} = 16^{\frac{1}{4}} = (2^4)^{\frac{1}{4}} = 2^{\frac{4}{4}} = 2$.

In particular, if $n = 2$ in $\sqrt[n]{A}$, we usually omit 2 in the radical sign $\sqrt[2]{}$, so we normally put it this way: $\sqrt{A}$, and call $\sqrt{}$ a square root sign. So $\sqrt{A}$ can be read as a square root of A.

And of course, the sign, $\sqrt{}$ can be called a second radical sign or a second root sign, too. So $\sqrt{A}$ can be read as a second radical of A or a second root of A, also.

So for instance, we can set: $\sqrt[2]{3} = \sqrt{3}$, which is usually called a square root of 3, and can be called a second radical of 3 or a second root of 3, too.

Besides, $\sqrt[3]{}$ is often called a cube root sign, so for instance, $\sqrt[3]{5}$ can be read as a cube root of 5. And of course, it can be called a third radical of 5 or a third root of 5, too.

And if $n \neq 2$ in $\sqrt[n]{A}$, we have to specify n, which indicates the degree of the radical. So for instance, we can set: $7^{\frac{1}{3}} = \sqrt[3]{7}$, which is a third root (radical) of 7, and can set: $3^{\frac{1}{7}} = \sqrt[7]{3}$, which is a seventh root (radical) of 3, etc.   What then, about $5^{\frac{2}{3}}$ or $7^{-\frac{2}{3}}$?

Let's first, express an nth root of a number in some more general manner.

Suppose that $A = b^m$, where $b \geq 0$, and m is an integer, and that n is an integer ≥ 2.

Then, we can get: $A = b^m \Rightarrow \sqrt[n]{A} = \sqrt[n]{b^m}$, which is an nth root in more general form.

And we can put it this way, too: $A = b^m \Rightarrow A^{\frac{1}{n}} = \sqrt[n]{A} = (b^m)^{\frac{1}{n}} = b^{\frac{m}{n}} \Rightarrow \sqrt[n]{A} = b^{\frac{m}{n}}$.

So we can put the nth root either of the two ways as follows: $\sqrt[n]{A} = \sqrt[n]{b^m}$, and $\sqrt[n]{A} = b^{\frac{m}{n}}$.

And thus, we can get: $b^{\frac{m}{n}} = \sqrt[n]{b^m}$, which can be read as 'an nth root of b to the m'.

So for instance, we can set: $5^{\frac{2}{3}} = \sqrt[3]{5^2}$, which is usually read as 'a cube root of five squared', and of course, can be read as 'a third root of five to the second', too.

And also, we can set: $7^{-\frac{2}{3}} = \sqrt[3]{7^{-2}} = \sqrt[3]{\frac{1}{7^2}} = ...$, or can set: $7^{-\frac{2}{3}} = \frac{1}{7^{\frac{2}{3}}} = \frac{1}{\sqrt[3]{7^2}} = ...$

And for more examples, using the exponential identities, we can get:

$$8^{\frac{2}{3}} = \sqrt[3]{8^2} = \sqrt[3]{(2^3)^2} = \sqrt[3]{(2^3)^2} = \sqrt[3]{2^6} = 2^{\frac{6}{3}} = 2^2 = 4.$$

$$8^{-\frac{2}{3}} = \sqrt[3]{8^{-2}} = \sqrt[3]{(8^{-1})^2} = \sqrt[3]{(\tfrac{1}{8})^2} = \sqrt[3]{(\tfrac{1}{2^3})^2} = \sqrt[3]{(\tfrac{1}{2})^6} = (\tfrac{1}{2})^{\frac{6}{3}} = 2^{-2} = \tfrac{1}{4}.$$

So we can put the same radical many different ways.

Let's next, take a look at some cases where the bases are fractions.

To begin with, we have an exponential identity where: $(\frac{x}{y})^z = \frac{x^z}{y^z}$, where both x and $y > 0$.

So assuming both a and $b > 0$, we can have:

$$\left(\frac{a}{b}\right)^{\frac{m}{n}} = \frac{a^{\frac{m}{n}}}{b^{\frac{m}{n}}} = \frac{\sqrt[n]{a^m}}{\sqrt[n]{b^m}} = \sqrt[n]{\frac{a^m}{b^m}} = \sqrt[n]{\left(\frac{a}{b}\right)^m}.$$ How come we can get: $\frac{\sqrt[n]{a^m}}{\sqrt[n]{b^m}} = \sqrt[n]{\frac{a^m}{b^m}}$, though?

We can have: $\frac{a^{\frac{m}{n}}}{b^{\frac{m}{n}}} = \left(\frac{a^m}{b^m}\right)^{\frac{1}{n}} = \sqrt[n]{\frac{a^m}{b^m}}$, and $\frac{a^{\frac{m}{n}}}{b^{\frac{m}{n}}} = \frac{\sqrt[n]{a^m}}{\sqrt[n]{b^m}}$. So we can get: $\frac{\sqrt[n]{a^m}}{\sqrt[n]{b^m}} = \sqrt[n]{\frac{a^m}{b^m}}$.

Now, what if $b < 0$ in $\sqrt[n]{b^m}$?

Then, $\sqrt[n]{b^m}$ might not be a real number.

For instance, $\sqrt[4]{(-16)^2}$ is a real number, but $\sqrt[4]{(-16)^3}$ is not a real number.   How come?

The number $\sqrt[4]{(-16)^3}$ means that multiplying 1 by $\sqrt[4]{(-16)^3}$ four times, we get $(-16)^3$.

However, $\sqrt[4]{(-16)^3}$ does not exist in the real number space, that is, it is not a real number.

That's because $(-16)^3$ is negative, and there is no real number by which we can multiply 1, four times to get $(-16)^3$.

In fact, no matter what real number it may be, multiplying 1 by it, even number of times, we get no negative number.

Let's take a closer look at the radical: $\sqrt[4]{(-16)^3}$.

We have: $(-16)^3 = (-2^4)^3 = -2^{12}$, so we get: $\sqrt[4]{(-16)^3} = \sqrt[4]{-2^{12}}$.

However, we have: $\sqrt[4]{-2^{12}} \neq -2^{\frac{12}{4}} = -2^3 = -8$. So we get: $\sqrt[4]{(-16)^3} \neq -8$.

That is to say that we get: $\sqrt[4]{(-16)^3} = (-16)^{\frac{3}{4}} = (-2^4)^{\frac{3}{4}} \neq -2^{4 \cdot \frac{3}{4}} = -2^3 = -8$.

• It's because processing a radical, we have to process what's inside the radical sign first.

So we cannot have: $\sqrt[4]{(-16)^3} = -2^{\frac{12}{4}}$, but we can have: $\sqrt[4]{(-16)^3} = \sqrt[4]{(-2^4)^3} = \sqrt[4]{-2^{12}}$, which is not however, a real number, because -2^{12} is negative, and no matter what real number it may be, multiplying 1 by it, even number of times, we cannot get a negative number.

Thus, in $\sqrt[n]{b^m}$, if the base $b < 0$, n is even, and m is odd, we get: $b^m < 0$, so $\sqrt[n]{b^m}$ cannot exist in the real number space, and thus, is not a real number.

In the real number space, no matter what number we may multiply 1 by, even number of times, we get no negative number, that is, we get a number ≥ 0.

However, if n is odd, $\sqrt[n]{b^m}$ is a real number even if the base $b < 0$.

That is to say that if n is odd, $\sqrt[n]{b^m}$ is a real number no matter what real number the base b may be.   How come?

To begin with, if $b = 0$, we just get: $\sqrt[n]{b^m} = 0$, which is of course, a real number.

Next, if $b > 0$, we simply get: $b^m > 0$, so $\sqrt[n]{b^m}$ is a real number regardless of n.

And next, if $b < 0$, we get: $b^m < 0$ if m is odd, so even if n is odd, $\sqrt[n]{b^m}$ is a real number, too.   How come?

There is a particular real number by which we can multiply 1, an odd number of times to get a number negative, and the particular real number is negative, too. For instance:

We have: $\sqrt[5]{-32} = (-32)^{\frac{1}{5}}$, which is a real number.   It's because:

A radical $\sqrt[5]{-32}$ means that multiplying 1 by the radical $\sqrt[5]{-32}$ itself, 5 times, we get –32.

And we have: $\sqrt[5]{-32} = \sqrt[5]{(-2)^5} = (-2)^{\frac{5}{5}} = (-2)^1 = -2$. So we get: $\sqrt[5]{-32} = -2$.

And thus, multiplying 1 by $\sqrt[5]{-32}$, 5 times, we multiply 1 by -2, 5 times, and then, we get: $(-2)^5 = -32$.

And for another instance, we have: $\sqrt[5]{(-32)^3} = (-32)^{\frac{3}{5}}$, which is real.   It's because:

$\sqrt[5]{(-32)^3}$ means that multiplying 1 by the radical $\sqrt[5]{(-32)^3}$ itself, 5 times, we get $(-32)^3$.

And we have: $\sqrt[5]{(-32)^3} = \sqrt[5]{\{(-2)^5\}^3} = \sqrt[5]{(-2)^{15}} = (-2)^{\frac{15}{5}} = (-2)^3 = -8$.

So we get: $\sqrt[5]{(-32)^3} = -8$.

And thus, multiplying 1 by $\sqrt[5]{(-32)^3}$, 5 times, we multiply 1 by -8, 5 times, and then we get: $(-8)^5 = \{(-2)^3\}^5 = (-2)^{15} = \{(-2)^5\}^3 = (-32)^3$.

So even if b is negative, $\sqrt[n]{b^m}$, that is, $b^{\frac{m}{n}}$ is a real number if m and n both are odd.

And note that we have: $-b^n \neq (-b)^n$, and that if $b = -2$ in $b^{\frac{m}{n}}$, we get: $(-2)^{\frac{m}{n}}$, and not: $-2^{\frac{m}{n}}$.

Also, if n and m both are even, $\sqrt[n]{b^m}$ is a real number even if the base $b < 0$ since $b^m > 0$.

For instance, we have: $\sqrt[5]{(-2)^2} = (-2)^{\frac{2}{5}} = \{(-2)^2\}^{\frac{1}{5}} = 4^{\frac{1}{5}}$, and $\sqrt[5]{(-2)^2} = \sqrt[5]{4}$, which is real.

Besides, when the *base* is *negative*, there can be a significant difference between using a radical sign and using power notation. And we will cover in the next section how it is the case. What in the world is the real number space, though?

It is a conceptual place where all real numbers exist, and real numbers only can exist. So the space of real numbers is in fact, a set of all real numbers, and therefore, if x does not exist in the real number space, x is not a real number.

So we can use the word, '*space*' for expressing an area or region where all the items in a particular category exist, and such items only can exist. And in such a space, the number of items is normally infinite. A space can mean therefore, an infinite set.

So for instance:

The space of integers or the integer space is a conceptual place or region where all integers exist, and integers only can exist. It is therefore, a set of all integers. So if A does not belong to an integer space, A is not an integer.

The space of rectangles is a set of all rectangles, and is a conceptual place where all rectangles exist, and rectangles only can exist. So if B does not belong to a space of rectangles, B is not a rectangle.

The space of functions is a conceptual place where all functions exist, and functions only can exist. So if C does not belong to the space of functions, C is not a function.

The space of sequences is a conceptual place where all sequences exist, and sequences only can exist. So if D does not exist in the space of sequences, D is not a sequence.

And note that an nth root $A^{\frac{1}{n}}$ is a power, too, because we can take A as the base, and can take $\frac{1}{n}$ as the exponent. So A in a radical $\sqrt[n]{A}$ is called the base of the radical, too.

That is to say that <u>what's inside a radical sign</u> is called <u>the base of the radical</u>.

So can the base of a radical be a power?

Yes, it can be. We can have a radical like this $\sqrt[n]{b^m}$, and in this case, the base of the radical is b^m, which is a power, so the base of a radical can be a power, too.
For instance, in a radical $\sqrt[4]{2^3}$, 2^3 is the base, and is a power of 2.

And next taking some samples on arithmetic in radicals, we can have:

$$2^2\sqrt{2} = 4\sqrt{2} = \sqrt{16}\sqrt{2} = \sqrt{16\cdot 2} = \sqrt{2^4 2} = \sqrt{2^5} = (2^5)^{\frac{1}{2}} = 2^{\frac{5}{2}}$$

$$2^2 2^{\frac{1}{2}} = 2^{2+\frac{1}{2}} = 2^{\frac{5}{2}}$$

$$2^2 \div \sqrt{2} = \frac{2^2}{\sqrt{2}} = \frac{4}{\sqrt{2}} = \frac{\sqrt{16}}{\sqrt{2}} = \sqrt{\frac{16}{2}} = \sqrt{8} = \sqrt{2^3} = (2^3)^{\frac{1}{2}} = 2^{\frac{3}{2}} = 2^{1+\frac{1}{2}} = 2^1 2^{\frac{1}{2}} = 2\sqrt{2}$$

$$2^2 \div 2^{\frac{1}{2}} = 2^{2-\frac{1}{2}} = 2^{\frac{3}{2}}$$

$$\frac{2^{10.2}}{2^{0.8}} = 2^{10.2-0.8} = 2^{9.4} = 2^{\frac{94}{100}} = 2^{\frac{47}{50}} = \sqrt[50]{2^{47}} \text{ , and also, } \frac{2^{10.2}}{2^{0.8}} = \frac{1}{2^{0.8-10.2}} = \frac{1}{2^{-9.4}} = 2^{9.4}$$

$$(2^2)^{\frac{1}{2}} = 2^{2\cdot\frac{1}{2}} = 2^1 = 2 \text{ , and also, } \sqrt{2^2} = (2^2)^{\frac{1}{2}} = 2 \text{ .}$$

$$6^{0.8} = 6^{\frac{8}{10}} = 6^{\frac{4}{5}} = \sqrt[5]{6^4} \text{ , and also, } 6^{0.8} = (2\cdot 3)^{0.8} = 2^{0.8} 3^{0.8} \text{ .}$$

$$\left(\frac{2}{3}\right)^{\frac{1}{2}} = \sqrt{\frac{2}{3}} = \frac{\sqrt{2}}{\sqrt{3}} = \frac{2^{\frac{1}{2}}}{3^{\frac{1}{2}}}, \ \left(\frac{2}{3}\right)^{\frac{2}{3}} = \sqrt[3]{\left(\frac{2}{3}\right)^2} = \sqrt[3]{\frac{2^2}{3^2}} = \frac{\sqrt[3]{2^2}}{\sqrt[3]{3^2}} = \frac{2^{\frac{2}{3}}}{3^{\frac{2}{3}}},$$

$$\left(\frac{2}{3}\right)^{-\frac{5}{3}} = \sqrt[3]{\left(\frac{2}{3}\right)^{-5}} = \sqrt[3]{\left(\frac{3}{2}\right)^5} = \sqrt[3]{\frac{3^5}{2^5}} = \frac{\sqrt[3]{3^5}}{\sqrt[3]{2^5}} = \frac{\sqrt[3]{3^{3+2}}}{\sqrt[3]{2^{3+2}}} = \frac{\sqrt[3]{3^3 3^2}}{\sqrt[3]{2^3 2^2}} = \frac{\sqrt[3]{3^3}\sqrt[3]{3^2}}{\sqrt[3]{2^3}\sqrt[3]{2^2}} = \frac{3\sqrt[3]{3^2}}{2\sqrt[3]{2^2}} = \frac{3^{\frac{5}{3}}}{2^{\frac{5}{3}}}$$

$$\left(\frac{2}{3}\right)^{-0.1} = \frac{2^{-0.1}}{3^{-0.1}} = \frac{\sqrt[10]{2^{-1}}}{\sqrt[10]{3^{-1}}} = \frac{\sqrt[10]{\frac{1}{2}}}{\sqrt[10]{\frac{1}{3}}} = \sqrt[10]{\frac{\frac{1}{2}}{\frac{1}{3}}} = \sqrt[10]{\frac{3}{2}}$$

$$\left(\frac{2}{3}\right)^{-0.1} = \left(\left(\frac{2}{3}\right)^{-1}\right)^{0.1} = \left(\frac{3}{2}\right)^{0.1} = \frac{3^{0.1}}{2^{0.1}} = \frac{\sqrt[10]{3}}{\sqrt[10]{2}} = \sqrt[10]{\frac{3}{2}}$$

And let's look at more closely, the case where the base b is 0 in a power b^p.

• Suppose first, the base b, and exponent p both are 0. Then, we get 0^0, which is technically a power, too, but cannot be one particular number.   How come?

We can get this: $0^0 = 1$, because multiplying 1 by 0 no times, we get 1, and we can get this, too: $0^0 = 0^{1-1} = \frac{0^1}{0^1} = \frac{0}{0}$. So 0^0 can be any number.   How come?

Setting: $x = 0^0$, we can get: $x = \frac{0}{0}$, because $\frac{0}{0} = 0^{1-1} = 0^0$.

So we can get: $0 \cdot x = 0$, since $x = \frac{0}{0}$.

Therefore, x can be any number, because $0 \cdot x = 0$ can hold for any value of x.
So x, that is, the power 0^0 cannot be one particular number.

Thus, the exponent p cannot be 0 if the base $b = 0$ in the power b^p.

• Suppose next, the base b is 0, and the exponent p is negative.

Then, we get such weird things as 0^{-1}, 0^{-3}, $0^{-0.2}$, etc., which are thus, all garbage in math. How come?

We know: $0^{-1} = \frac{1}{0}$, $0^{-3} = \frac{1}{0^3} = \frac{1}{0}$, and $0^{-0.2} = \frac{1}{0^{0.2}} = \frac{1}{0}$, but no division by 0 is allowed.

Consequently therefore, if the base b is 0 in a power b^p, the exponent p cannot be ≤ 0, that is, p has to be > 0 only.

So if we want to use all real numbers as the exponent p, we don't want to use 0 as the base b. And the same is true for the case where the base b is negative, too. How come?

We will cover it in the next three sections after taking some sets of examples in radicals in the next pages.

Examples 1 in Radicals

Simplify each radical below or put each in a mixed radical if it can be done so. A mixed radical is a radical multiplied by a rational such as $2\sqrt{3}$ and $4\sqrt[3]{2}$.

0. $\sqrt{4}$ 1. $\sqrt{8}$ 2. $\sqrt{-2}$ 3. $\sqrt{12}$ 4. $\sqrt[3]{8}$

5. $\sqrt[3]{32}$ 6. $\sqrt[3]{-8}$ 7. $\sqrt{1.69}$ 8. $\sqrt[3]{5.6}$ 9. $\sqrt[3]{0.016}$

A. $\sqrt[3]{-3.2}$ B. $\sqrt[3]{270}$ C. $\sqrt[3]{0.081}$ D. $\sqrt[3]{-16}$

60

Suggestions or Solutions
To the Examples 1 in Radicals

0. $\sqrt{4} = \sqrt{2^2} = 2$, because $1 \times \sqrt{4} \times \sqrt{4} = 4$, which is what's inside the square root sign, and we have: $2^2 = 4$, too, which means we have: $1 \cdot 2 \cdot 2 = 4$. So we get: $\sqrt{4} = 2$.
What then, about: $1 \cdot (-2) \cdot (-2) = 4$?

We know $\sqrt{4}$ is positive. So we don't get this: $\sqrt{4} = -2$.

And by the same token, we get: $\sqrt{9} = \sqrt{3^2} = 3$, $\sqrt{16} = \sqrt{4^2} = 4$, $\sqrt{25} = \sqrt{5^2} = 5$, etc.

1. $\sqrt{8} = \sqrt{4 \cdot 2} = \sqrt{4}\sqrt{2} = 2\sqrt{2}$. How come though, do we get: $\sqrt{8} = \sqrt{4 \cdot 2} = \sqrt{4}\sqrt{2}$?

By definition, we get: $1 \times \sqrt{8} \times \sqrt{8} = 8$.
Andy by definition again, we get: $1 \times \sqrt{4 \cdot 2} \times \sqrt{4 \cdot 2} = 4 \cdot 2 = 8$.
And also, we get: $1 \times \sqrt{4}\sqrt{2} \times \sqrt{4}\sqrt{2} = \sqrt{4} \times \sqrt{4} \times \sqrt{2} \times \sqrt{2} = 4 \times 2 = 8$.
So we get: $\sqrt{8} = \sqrt{4 \cdot 2} = \sqrt{4}\sqrt{2}$.

And by the same token, we get:

$$\sqrt{21} = \sqrt{3 \cdot 7} = \sqrt{3}\sqrt{7}, \quad \sqrt{15} = \sqrt{3 \cdot 5} = \sqrt{5}\sqrt{3}, \quad \sqrt{54} = \sqrt{6 \cdot 9} = \sqrt{9}\sqrt{6} = 3\sqrt{6}$$

$$\sqrt{21} = \sqrt{3 \cdot 7} = \sqrt{3}\sqrt{7}, \quad \sqrt{24} = \sqrt{6 \cdot 4} = \sqrt{6}\sqrt{4} = 2\sqrt{6}, \quad \sqrt{72} = \sqrt{36 \cdot 2} = \sqrt{36}\sqrt{2} = 6\sqrt{2}$$

2. $\sqrt{-2}$ cannot be made within the set of all real numbers, and thus, is not allowed, because no real number can be squared to be a negative number.

3. $\sqrt{12} = \sqrt{4 \cdot 3} = \sqrt{4}\sqrt{3} = 2\sqrt{3}$ And by the same token, we get:

$\sqrt{18} = \sqrt{9 \cdot 2} = \sqrt{9}\sqrt{2} = 3\sqrt{2}$, $\sqrt{50} = \sqrt{25 \cdot 2} = 5\sqrt{2}$, $\sqrt{48} = \sqrt{16 \cdot 3} = 4\sqrt{3}$

$\sqrt{242} = \sqrt{121 \cdot 2} = \sqrt{11^2}\sqrt{2} = 11\sqrt{2}$, $\sqrt{432} = \sqrt{144 \cdot 3} = \sqrt{12^2}\sqrt{3} = 12\sqrt{3}$, etc.

4. $\sqrt[3]{8} = \sqrt[3]{2^3} = 2$, because $1 \times \sqrt[3]{8} \times \sqrt[3]{8} \times \sqrt[3]{8} = 8$, which is what's inside the cube root sign, and we have: $2^3 = 8$, too, which means we have: $\mathbf{1 \cdot 2 \cdot 2 \cdot 2 = 8}$. So we get: $\sqrt[3]{8} = \mathbf{2}$.

And by the same token, we get:
$\sqrt[3]{27} = \sqrt[3]{3^3} = 3$, $\sqrt[3]{64} = \sqrt[3]{4^3} = 4$, $\sqrt[3]{125} = \sqrt[3]{5^3} = 5$, $\sqrt[3]{216} = \sqrt[3]{6^3} = 6$, etc.

5. $\sqrt[3]{32} = \sqrt[3]{8 \cdot 4} = \sqrt[3]{8} \cdot \sqrt[3]{4} = 2\sqrt[3]{4}$ And by the same token, we get:

$\sqrt[3]{40} = \sqrt[3]{5 \cdot 8} = \sqrt[3]{5} \cdot \sqrt[3]{8} = 2\sqrt[3]{5}$, $\sqrt[3]{54} = \sqrt[3]{27 \cdot 2} = \sqrt[3]{27} \cdot \sqrt[3]{2} = 3\sqrt[3]{2}$

$\sqrt[3]{72} = \sqrt[3]{9 \cdot 8} = \sqrt[3]{8} \cdot \sqrt[3]{9} = 2\sqrt[3]{9}$, $\sqrt[3]{81} = \sqrt[3]{27 \cdot 3} = \sqrt[3]{27} \cdot \sqrt[3]{3} = 3\sqrt[3]{3}$, $\sqrt[3]{162} = \sqrt[3]{27 \cdot 6} = 3\sqrt[3]{6}$

6. $\sqrt[3]{-8} = \sqrt[3]{(-2)^3} = -2$, because $1 \times \sqrt[3]{-8} \times \sqrt[3]{-8} \times \sqrt[3]{-8} = -8$, which is what's inside the cube root sign, and we have: $(-2)^3 = -8$, too, which means we have: $\mathbf{1 \cdot (-2) \cdot (-2) \cdot (-2) = -8}$.

So we get: $\sqrt[3]{-8} = \mathbf{-2}$.   And by the same token, we get:

$\sqrt[3]{-27} = \sqrt[3]{(-3)^3} = -3$, $\sqrt[3]{-64} = \sqrt[3]{(-4)^3} = -4$, $\sqrt[3]{-0.008} = \sqrt[3]{(-0.2)^3} = -0.2$, etc.

7. $\sqrt{1.69} = \sqrt{(1.3)^2} = 1.3$

8. $\sqrt[3]{5.6} = \sqrt[3]{8 \cdot 0.7} = \sqrt[3]{8} \cdot \sqrt[3]{0.7} = 2\sqrt[3]{0.7}$

9. $\sqrt[3]{0.016} = \sqrt[3]{8 \cdot 0.002} = \sqrt[3]{8} \cdot \sqrt[3]{0.001} \cdot \sqrt[3]{2} = 2 \cdot 0.1\sqrt[3]{2} = 0.2\sqrt[3]{2}$

A. $\sqrt[3]{-3.2} = \sqrt[3]{-8 \cdot 0.4} = \sqrt[3]{-8} \cdot \sqrt[3]{0.4} = -2\sqrt[3]{0.4}$

B. $\sqrt[3]{270} = \sqrt[3]{27 \cdot 10} = \sqrt[3]{27}\sqrt[3]{10} = 3\sqrt[3]{10}$

C. $\sqrt[3]{0.081} = \sqrt[3]{0.001 \cdot 81} = \sqrt[3]{0.001}\sqrt[3]{81} = 0.1\sqrt[3]{27 \cdot 3} = 0.1\sqrt[3]{3^3 3} = 0.3\sqrt[3]{3}$

D. $\sqrt[3]{-160} = \sqrt[3]{-8 \cdot 20} = \sqrt[3]{-2^3 20} = -2\sqrt[3]{20}$

Note that: $-2^3 = (-2)^3 = -8$, but $-2^2 \neq (-2)^2$, because $-2^2 = -4$, and $(-2)^2 = 4$.

Examples 2 in Radicals

0. $\sqrt{3} \times \sqrt{7}$

1. $\sqrt[3]{3} \times \sqrt[3]{4}$

2. $\sqrt{4}\sqrt{8}$

3. $\sqrt{18}\sqrt{8}$

4. $\sqrt{45}\sqrt{20}$

5. $\sqrt{24}\sqrt{14}$

6. $\sqrt{21}\sqrt{45}$

7. $\sqrt{150}\sqrt{28}$

8. $\sqrt{72}\sqrt{36}$

9. $\sqrt{27}\sqrt{24}$

A. $\sqrt{54}\sqrt{162}$

B. $\sqrt{625}\sqrt{125}$

C. $\sqrt{48}\sqrt{216}$

D. $\sqrt{363}\sqrt{338}$

E. $\sqrt{150}\sqrt{432}$

F. $\sqrt[3]{54}\sqrt[3]{192}$

G. $\sqrt[3]{250}\sqrt[3]{432}$

H. $\sqrt[3]{108}\sqrt[3]{32}$

I. $\sqrt[3]{72}\sqrt[3]{243}$

J. $\sqrt[3]{-54}\sqrt[3]{-128}$

K. $\sqrt[3]{-0.008} \cdot \sqrt[3]{125}$

L. $\sqrt[3]{1.6} \cdot \sqrt[3]{16}$

M. $\sqrt[3]{270}\sqrt[3]{-80}$

N. $\sqrt[3]{0.016} \cdot \sqrt[3]{16}$

Suggestions or Solutions
To the Examples 2 in Radicals

0. $\sqrt{3} \times \sqrt{7} = \sqrt{3} \cdot \sqrt{7} = \sqrt{3}\sqrt{7} = \sqrt{3 \times 7} = \sqrt{3 \cdot 7} = \sqrt{21}$

1. $\sqrt[3]{3} \times \sqrt[3]{4} = \sqrt[3]{3} \cdot \sqrt[3]{4} = \sqrt[3]{3}\sqrt[3]{4} = \sqrt[3]{3 \times 4} = \sqrt[3]{3 \cdot 4} = \sqrt[3]{12}$

2. $\sqrt{4}\sqrt{8} = 2 \cdot 2\sqrt{2} = 4\sqrt{2}$, or $\sqrt{4}\sqrt{8} = \sqrt{32} = \sqrt{16 \cdot 2} = 4\sqrt{2}$

3. $\sqrt{18}\sqrt{8} = 3\sqrt{2} \cdot 2\sqrt{2} = 6 \cdot 2 = 12$, or $\sqrt{18}\sqrt{8} = \sqrt{144} = \sqrt{12^2} = 12$

4. $\sqrt{45}\sqrt{20} = 3\sqrt{5} \cdot 2\sqrt{5} = 6 \cdot 5 = 30$, or $\sqrt{45}\sqrt{20} = \sqrt{900} = \sqrt{30^2} = 30$

5. $\sqrt{24}\sqrt{14} = 2\sqrt{6} \cdot \sqrt{14} = 2\sqrt{6 \cdot 14} = 2\sqrt{2 \cdot 3 \cdot 2 \cdot 7} = 2\sqrt{4 \cdot 3 \cdot 7} = 4\sqrt{3 \cdot 7} = 4\sqrt{21}$

6. $\sqrt{21}\sqrt{45} = \sqrt{3 \cdot 7} \cdot \sqrt{5 \cdot 9} = 3\sqrt{3 \cdot 5 \cdot 7} = 3\sqrt{105}$

7. $\sqrt{150}\sqrt{28} = \sqrt{25 \cdot 6}\sqrt{7 \cdot 4} = 5\sqrt{6} \cdot 2\sqrt{7} = 10\sqrt{6 \cdot 7} = 10\sqrt{42}$

8. $\sqrt{72}\sqrt{36} = \sqrt{36 \cdot 2}\sqrt{36} = 36\sqrt{2}$

9. $\sqrt{27}\sqrt{24} = \sqrt{9 \cdot 3}\sqrt{4 \cdot 6} = 3\sqrt{3} \cdot 2\sqrt{3 \cdot 2} = 3 \cdot 2 \cdot \sqrt{3}\sqrt{3 \cdot 2} = 6 \cdot 3\sqrt{2} = 18\sqrt{2}$

A. $\sqrt{54}\sqrt{162} = \sqrt{9\cdot6}\sqrt{81\cdot2} = 3\sqrt{6}\cdot9\sqrt{2} = 18\sqrt{12} = 18\sqrt{4\cdot3} = 36\sqrt{3}$,

B. $\sqrt{625}\sqrt{125} = \sqrt{25^2}\sqrt{25\cdot5} = 25\cdot5\sqrt{5} = 125\sqrt{5}$

In fact, $625 = 125\cdot5$, so $\sqrt{625}\sqrt{125} = \sqrt{125\cdot5}\sqrt{125} = 125\sqrt{5}$

C. $\sqrt{48}\sqrt{216} = \sqrt{16\cdot3}\sqrt{36\cdot6} = 4\sqrt{3}\cdot6\sqrt{6} = 24\cdot3\sqrt{2} = 72\sqrt{2}$

D. $\sqrt{363}\sqrt{338} = \sqrt{121\cdot3}\sqrt{169\cdot2} = \sqrt{11^2}\sqrt{3}\sqrt{13^2}\sqrt{2} = 11\cdot13\sqrt{6} = 143\sqrt{6}$

E. $\sqrt{150}\sqrt{432} = \sqrt{50\cdot3}\sqrt{144\cdot3} = \sqrt{25\cdot6}\sqrt{12^2}\sqrt{3} = 5\sqrt{6}\cdot12\sqrt{3} = 60\cdot3\sqrt{2} = 180\sqrt{2}$

F. $\sqrt[3]{54}\sqrt[3]{192} = \sqrt[3]{27\cdot2}\sqrt[3]{64\cdot3} = 3\sqrt[3]{2}\cdot4\sqrt[3]{3} = 12\sqrt[3]{2\cdot3} = 12\sqrt[3]{6}$

G. $\sqrt[3]{250}\sqrt[3]{432} = \sqrt[3]{125\cdot2}\sqrt[3]{216\cdot2} = \sqrt[3]{5^3 2}\sqrt[3]{6^3 2} = 5\sqrt[3]{2}\cdot6\sqrt[3]{2} = 30\sqrt[3]{4}$

H. $\sqrt[3]{108}\sqrt[3]{32} = \sqrt[3]{27\cdot4}\sqrt[3]{8\cdot4} = 3\sqrt[3]{4}\cdot2\sqrt[3]{4} = 8\sqrt[3]{16} = 8\sqrt[3]{8\cdot2} = 16\sqrt[3]{2}$

I. $\sqrt[3]{72}\sqrt[3]{243} = \sqrt[3]{9\cdot8}\sqrt[3]{27\cdot9} = 2\sqrt[3]{9}\cdot3\sqrt[3]{9} = 6\sqrt[3]{81} = 6\sqrt[3]{27\cdot3} = 18\sqrt[3]{3}$

J. $\sqrt[3]{-54}\sqrt[3]{-128} = \sqrt[3]{-27\cdot2}\sqrt[3]{-64\cdot2} = -3\sqrt[3]{2}(-4\sqrt[3]{2}) = 12\sqrt[3]{4}$

K. $\sqrt[3]{-0.008} \cdot \sqrt[3]{125} = \sqrt[3]{(-0.2)^3} \sqrt{5^3} = -0.2 \cdot 5 = -1$

L. $\sqrt[3]{1.6} \cdot \sqrt[3]{16} = \sqrt[3]{8 \cdot 0.2} \cdot \sqrt[3]{8 \cdot 2} = 2\sqrt[3]{0.2} \cdot 2\sqrt[3]{2} = 4\sqrt[3]{0.4}$

M. $\sqrt[3]{270}\sqrt[3]{-80} = \sqrt[3]{27 \cdot 10} \cdot \sqrt[3]{-8 \cdot 10} = 3\sqrt[3]{10}(-2\sqrt[3]{10}) = -6\sqrt[3]{100}$

N. $\sqrt[3]{0.016} \cdot \sqrt[3]{16} = \sqrt[3]{0.001 \cdot 16} \cdot \sqrt[3]{16} = 0.1\sqrt[3]{16} \cdot \sqrt[3]{16} = 0.1\sqrt[3]{16 \cdot 16}$

$= 0.1\sqrt[3]{64 \cdot 4} = 0.1\sqrt[3]{64}\sqrt[3]{4} = 0.1 \cdot 4\sqrt[3]{4} = 0.4\sqrt[3]{4}$

Examples 3 in Radicals

Simplify each radical below or put each in a mixed radical if it can be done so. A mixed radical is a radical multiplied by a rational such as $2\sqrt{3}$ and $4\sqrt[3]{2}$.

0. $\sqrt{\dfrac{1}{4}}$ 　　　 1. $\sqrt{\dfrac{1}{8}}$ 　　　 2. $\sqrt{\dfrac{-3}{8}}$ 　　　 3. $\sqrt[3]{\dfrac{8}{-27}}$

4. $\sqrt[3]{-\dfrac{8}{49}}$ 　　 5. $\sqrt{\dfrac{7}{8}}$ 　　 6. $\sqrt{\dfrac{3}{-4}}$ 　　 7. $\sqrt[3]{\dfrac{2}{27}}$

8. $\sqrt[3]{-\dfrac{16}{27}}$ 　 9. $\sqrt{\dfrac{4}{27}}$ 　 A. $\sqrt[3]{\dfrac{27}{4}}$ 　 B. $\sqrt[3]{\dfrac{108}{27}}$

C. $\sqrt{\dfrac{45}{98}}$ 　 D. $\sqrt[3]{\dfrac{1}{4}}$ 　 E. $\sqrt[3]{\dfrac{1}{24}}$ 　 F. $\sqrt[3]{\dfrac{3}{8}}$

G. $\sqrt[3]{\dfrac{4}{9}}$ 　 H. $\sqrt{\dfrac{4}{9}}$ 　 I. $\sqrt{\dfrac{2}{9}}$ 　 J. $\sqrt{-\dfrac{4}{49}}$

K. $\sqrt[3]{\dfrac{15}{18}}$ 　 L. $\sqrt[3]{\dfrac{-1}{24}}$ 　 M. $\sqrt{\dfrac{8}{3}}$ 　 N. $\sqrt[3]{\dfrac{8}{81}}$

O. $\sqrt[3]{\dfrac{16}{9}}$ 　 P. $\sqrt[3]{\dfrac{135}{98}}$

Suggestions or Solutions
To the Examples 3 in Radicals

0. $\sqrt{\frac{1}{4}} = \sqrt{\left(\frac{1}{2}\right)^2} = \frac{1}{2}$, and we can put it this way, too, of course: $\sqrt{\frac{1}{4}} = \sqrt{\frac{1}{2^2}} = \frac{\sqrt{1}}{\sqrt{2^2}} = \frac{1}{2}$.

1. $\sqrt{\frac{1}{8}} = \sqrt{\frac{1}{4} \cdot \frac{1}{2}} = \sqrt{\left(\frac{1}{2}\right)^2 \frac{1}{2}} = \frac{1}{2}\sqrt{\frac{1}{2}}$, and this way, too: $\sqrt{\frac{1}{8}} = \sqrt{\left(\frac{1}{2}\right)^3} = \sqrt{\left(\frac{1}{2}\right)^2 \frac{1}{2}} = \frac{1}{2}\sqrt{\frac{1}{2}}$.

2. $\sqrt{\frac{-3}{8}}$ cannot be made within the set of all real numbers, and thus, is not allowed, because no real number can be squared to be a negative number.

3. $\sqrt[3]{\frac{8}{-27}} = \sqrt[3]{-\frac{8}{27}} = -\sqrt[3]{\frac{2^3}{3^3}} = -\sqrt[3]{\left(\frac{2}{3}\right)^3} = -\frac{2}{3}$

4. $\sqrt[3]{-\frac{8}{49}} = -\sqrt[3]{\frac{8}{49}} = -\sqrt[3]{\frac{2^3}{49}} = -\frac{\sqrt[3]{2^3}}{\sqrt[3]{49}} = -\frac{2}{\sqrt[3]{49}}$, which can be put the way below, too:

$-\frac{2}{\sqrt[3]{49}} = -\frac{2\sqrt[3]{7}}{\sqrt[3]{49}\sqrt[3]{7}} = -\frac{2\sqrt[3]{7}}{\sqrt[3]{49 \cdot 7}} = -\frac{2\sqrt[3]{7}}{\sqrt[3]{7^3}} = -\frac{2\sqrt[3]{7}}{7}$

5. $\sqrt{\frac{7}{8}} = \sqrt{\frac{7}{4 \cdot 2}} = \sqrt{\frac{1}{4} \cdot \frac{7}{2}} = \sqrt{\left(\frac{1}{2}\right)^2 \frac{7}{2}} = \frac{1}{2}\sqrt{\frac{7}{2}}$

6. $\sqrt{\frac{3}{-4}}$ cannot be made within the set of all real numbers, and thus, is not allowed, because no real number can be squared to be a negative number.

7. $\sqrt[3]{\frac{2}{27}} = \sqrt[3]{\frac{2}{3^3}} = \sqrt[3]{\frac{1}{3^3}\cdot 2} = \sqrt[3]{\left(\frac{1}{3}\right)^3 2} = \sqrt[3]{\left(\frac{1}{3}\right)^3}\cdot\sqrt[3]{2} = \frac{1}{3}\sqrt[3]{2}$, which equals: $\frac{\sqrt[3]{2}}{3}$.

8. $\sqrt[3]{-\frac{16}{27}} = -\sqrt[3]{\frac{8\cdot 2}{3^3}} = -\frac{1}{3}\sqrt[3]{8\cdot 2} = -\frac{1}{3}\sqrt[3]{2^3\cdot 2} = -\frac{1}{3}\sqrt[3]{2^3}\cdot\sqrt[3]{2} = -\frac{2}{3}\sqrt[3]{2}$, which equals: $-\frac{2\sqrt[3]{2}}{3}$.

9. $\sqrt{\frac{4}{27}} = \sqrt{\frac{2^2}{3^2\cdot 3}} = \sqrt{\frac{2^2}{3^2}\cdot\frac{1}{3}} = \sqrt{\left(\frac{2}{3}\right)^2\frac{1}{3}} = \frac{2}{3}\sqrt{\frac{1}{3}} = \frac{2}{3}\cdot\frac{1}{\sqrt3} = \frac{2}{3\sqrt3} = \frac{2\sqrt3}{3\sqrt3\sqrt3} = \frac{2\sqrt3}{3\cdot3} = \frac{2\sqrt3}{9}$

A. $\sqrt[3]{\frac{27}{4}} = \sqrt[3]{\frac{3^3}{4}} = \frac{\sqrt[3]{3^3}}{\sqrt[3]{4}} = \frac{3}{\sqrt[3]{4}} = \frac{3\sqrt[3]{2}}{\sqrt[3]{4}\cdot\sqrt[3]{2}} = \frac{3\sqrt[3]{2}}{\sqrt[3]{4\cdot2}} = \frac{3\sqrt[3]{2}}{\sqrt[3]{2^3}} = \frac{3\sqrt[3]{2}}{2} = \frac{3}{2}\sqrt[3]{2}$

B. $\sqrt[3]{\frac{108}{27}} = \sqrt[3]{\frac{27\cdot4}{3^3}} = \sqrt[3]{\frac{3^3\cdot4}{3^3}} = \sqrt[3]{4}$

C. $\sqrt{\frac{45}{98}} = \sqrt{\frac{9\cdot5}{49\cdot2}} = \sqrt{\frac{3^2\cdot5}{7^2\cdot2}} = \sqrt{\frac{3^2}{7^2}}\sqrt{\frac{5}{2}} = \frac{3}{7}\sqrt{\frac{5}{2}} = \frac{3}{7}\frac{\sqrt5}{\sqrt2} = \frac{3}{7}\frac{\sqrt5\sqrt2}{\sqrt2\sqrt2} = \frac{3}{7}\frac{\sqrt{5\cdot2}}{2} = \frac{3\sqrt{10}}{14}$

D. $\sqrt[3]{\frac{1}{4}}$ does not really have to be simplified, but can be changed to $\sqrt[3]{\frac{2}{8}} = \frac{\sqrt[3]{2}}{\sqrt[3]{8}} = \frac{\sqrt[3]{2}}{2}$.

E. $\sqrt[3]{\frac{1}{24}} = \sqrt[3]{\frac{1}{8\cdot3}} = \sqrt[3]{\frac{1}{8}}\cdot\sqrt[3]{\frac{1}{3}} = \frac{1}{2}\sqrt[3]{\frac{1}{3}} = \frac{1}{2}\frac{1}{\sqrt[3]{3}} = \frac{1}{2\sqrt[3]{3}} = \frac{\sqrt[3]{3^2}}{2\sqrt[3]{3}\cdot\sqrt[3]{3^2}} = \frac{\sqrt[3]{3^2}}{2\cdot3} = \frac{\sqrt[3]{9}}{6}$

F. $\sqrt[3]{\frac{3}{8}} = \sqrt[3]{\frac{3}{2^3}} = \frac{\sqrt[3]{3}}{\sqrt[3]{2^3}} = \frac{\sqrt[3]{3}}{2}$

G. $\sqrt[3]{\frac{4}{9}} = \sqrt[3]{\frac{12}{27}} = \frac{\sqrt[3]{12}}{\sqrt[3]{27}} = \frac{\sqrt[3]{12}}{3}$

H. $\sqrt{\frac{4}{9}} = \sqrt{\left(\frac{2}{3}\right)^2} = \frac{2}{3}$

I. $\sqrt{\frac{2}{9}} = \frac{\sqrt{2}}{\sqrt{9}} = \frac{\sqrt{2}}{3}$

J. $\sqrt{-\frac{4}{49}}$ cannot be made within the set of all real numbers, and thus, is not allowed, because no real number can be squared to be a negative number.

K. $\sqrt[3]{\frac{15}{18}} = \sqrt[3]{\frac{5}{6}}$, which does not really have to be simplified, but can be changed to

$\sqrt[3]{\frac{5 \cdot 36}{6 \cdot 36}} = \frac{\sqrt[3]{180}}{6}$.

L. $\sqrt[3]{\frac{-1}{24}} = -\sqrt[3]{\frac{1}{24}} = -\sqrt[3]{\frac{1}{8 \cdot 3}} = -\sqrt[3]{\frac{1}{8}} \cdot \sqrt[3]{\frac{1}{3}} = -\frac{1}{2}\sqrt[3]{\frac{9}{27}} = -\frac{1}{2}\frac{\sqrt[3]{9}}{\sqrt[3]{27}} = -\frac{1}{2}\frac{\sqrt[3]{9}}{3} = -\frac{\sqrt[3]{9}}{6}$

M. $\sqrt{\frac{8}{3}} = \sqrt{\frac{4 \cdot 2}{3}} = 2\sqrt{\frac{2}{3}} = 2\frac{\sqrt{2}}{\sqrt{3}} = \frac{2\sqrt{2}}{\sqrt{3}} = \frac{2\sqrt{2}\sqrt{3}}{\sqrt{3}\sqrt{3}} = \frac{2\sqrt{6}}{3}$

N. $\sqrt[3]{\frac{8}{81}} = \frac{\sqrt[3]{8}}{\sqrt[3]{81}} = \frac{2\sqrt[3]{9}}{\sqrt[3]{81 \cdot 9}} = \frac{2\sqrt[3]{9}}{9}$

O. $\sqrt[3]{\frac{16}{9}} = \sqrt[3]{\frac{16 \cdot 3}{27}} = \frac{\sqrt[3]{48}}{\sqrt[3]{27}} = \frac{\sqrt[3]{48}}{3}$

P. $\sqrt[3]{\frac{135}{98}} = \sqrt[3]{\frac{27 \cdot 5}{49 \cdot 2}} = \sqrt[3]{\frac{3^3 \cdot 5 \cdot 7}{7^3 \cdot 2}} = \frac{3}{7}\sqrt[3]{\frac{5 \cdot 7}{2}} = \frac{3}{7}\sqrt[3]{\frac{5 \cdot 7 \cdot 4}{2^3}} = \frac{3}{14}\sqrt[3]{140} = \frac{3\sqrt[3]{140}}{14}$

Examples 4 in Radicals

Simplify each radical below or put each in a mixed radical if it can be done so. A mixed radical is a radical multiplied by a rational such as $2\sqrt{3}$ and $4\sqrt[3]{2}$.

0. $\sqrt{2\sqrt{4}}$ 1. $\sqrt{2\sqrt{16}}$ 2. $\sqrt{3\sqrt{36}}$ 3. $\sqrt{6\sqrt{225}}$

4. $\sqrt{2\sqrt[3]{125}}$ 5. $\sqrt[3]{4\sqrt{9}}$ 6. $\sqrt[3]{4\sqrt{4}}$ 7. $\sqrt[3]{32\sqrt{16}}$

8. $\sqrt[3]{10\sqrt[3]{64}}$ 9. $\sqrt[3]{0.8}$ A. $\sqrt[3]{0.008}$ B. $\sqrt[3]{0.016}$

C. $\sqrt[3]{\dfrac{0.001}{0.027}}$ D. $\sqrt[3]{\dfrac{0.01}{2.7}}$ E. $\sqrt{\dfrac{4.84}{9}}$ F. $\sqrt{\dfrac{0.0004}{0.81}}$

G. $\sqrt[3]{\dfrac{-0.001}{0.027}}$ H. $\sqrt{\dfrac{0.4}{0.009}}$ I. $\sqrt[3]{\dfrac{8}{0.081}}$ J. $\sqrt[3]{\dfrac{1.5}{9}}$

Suggestions or Solutions
To the Examples 4 in Radicals

0. $\sqrt{2\sqrt{4}} = \sqrt{2\sqrt{2^2}} = \sqrt{2\cdot 2} = 2$. So by the same token, we can have:

$\sqrt{6\sqrt{3\sqrt{4\sqrt{9}}}} = \sqrt{6\sqrt{3\sqrt{4\sqrt{3^2}}}} = \sqrt{6\sqrt{3\sqrt{4\cdot 3}}} = \sqrt{6\sqrt{3\sqrt{12}}} = \sqrt{6\sqrt{36}} = \sqrt{6^2} = 6$.

1. $\sqrt{2\sqrt{16}} = \sqrt{2\sqrt{4^2}} = \sqrt{2\cdot 4} = \sqrt{2}\sqrt{4} = 2\sqrt{2}$

2. $\sqrt{3\sqrt{36}} = \sqrt{3\cdot 6} = \sqrt{3\cdot 3\cdot 2} = 3\sqrt{2}$

3. $\sqrt{6\sqrt{225}} = \sqrt{6\sqrt{25\cdot 9}} = \sqrt{6\sqrt{15\cdot 15}} = \sqrt{6\cdot 15} = \sqrt{3\cdot 2\cdot 3\cdot 5} = 3\sqrt{10}$

4. $\sqrt{2\sqrt[3]{125}} = \sqrt{2\sqrt[3]{5^3}} = \sqrt{2\cdot 5} = \sqrt{10}$

5. $\sqrt[3]{4\sqrt{9}} = \sqrt[3]{4\cdot 3} = \sqrt[3]{12}$

6. $\sqrt[3]{4\sqrt{4}} = \sqrt[3]{4\cdot 2} = 2$

7. $\sqrt[3]{32\sqrt{16}} = \sqrt[3]{8\cdot 4\cdot 4} = \sqrt[3]{8\cdot 8\cdot 2} = \sqrt[3]{8}\cdot\sqrt[3]{8}\cdot\sqrt[3]{2} = 2\cdot 2\sqrt[3]{2} = 4\sqrt[3]{2}$

8. $\sqrt[3]{10\sqrt[3]{64}} = \sqrt[3]{10\sqrt[3]{4^3}} = \sqrt[3]{10\cdot 4} = \sqrt[3]{5\cdot 8} = 2\sqrt[3]{5}$

9. $\sqrt{0.8} = \sqrt{\frac{8}{10}} = \sqrt{\frac{4}{5}} = \frac{\sqrt{4}}{\sqrt{5}} = \frac{2}{\sqrt{5}} = \frac{2\sqrt{5}}{\sqrt{5}\sqrt{5}} = \frac{2\sqrt{5}}{5}$

A. $\sqrt[3]{0.008} = \sqrt[3]{0.001 \cdot 8} = \sqrt[3]{(0.1)^3 \cdot 8} = \sqrt[3]{(0.1)^3} \cdot \sqrt[3]{8} = 0.1 \cdot 2 = 0.2.$

And of course, we can put it this way, too: $\sqrt[3]{0.008} = \sqrt[3]{(0.2)^3} = 0.2.$

B. $\sqrt[3]{0.016} = \sqrt[3]{\frac{16}{1000}} = \sqrt[3]{\frac{8 \cdot 2}{1000}} = \sqrt[3]{\frac{8}{1000} \cdot 2} = \sqrt[3]{(\frac{2}{10})^3 \cdot 2} = \sqrt[3]{(\frac{2}{10})^3} \cdot \sqrt[3]{2} = \frac{2}{10} \cdot \sqrt[3]{2} = \frac{1}{5} \cdot \sqrt[3]{2} = \frac{\sqrt[3]{2}}{5}$

C. $\sqrt[3]{\frac{0.001}{0.027}} = \sqrt[3]{\frac{1}{27}} = \frac{1}{3}.$ And we can put it this way, too: $\sqrt[3]{\frac{0.001}{0.027}} = \sqrt[3]{(\frac{0.1}{0.3})^3} = \sqrt[3]{(\frac{1}{3})^3} = \frac{1}{3}.$

D. $\sqrt[3]{\frac{0.01}{2.7}} = \sqrt[3]{\frac{0.1}{27}} = \sqrt[3]{\frac{1}{27} \cdot 0.1} = \sqrt[3]{\frac{1}{27}} \cdot \sqrt[3]{0.1} = \frac{1}{3}\sqrt[3]{\frac{1}{10}} = \frac{1}{3} \cdot \frac{1}{\sqrt[3]{10}} = \frac{1}{3} \cdot \frac{\sqrt[3]{100}}{\sqrt[3]{10} \cdot \sqrt[3]{100}} = \frac{1}{3} \cdot \frac{\sqrt[3]{100}}{10} = \frac{\sqrt[3]{100}}{30}$

E. $\sqrt{\frac{4.84}{9}} = \sqrt{\frac{484}{900}} = \sqrt{\frac{121 \cdot 4}{900}} = \sqrt{\frac{121 \cdot 4}{225 \cdot 4}} = \sqrt{\frac{121}{225}} = \sqrt{(\frac{11}{15})^2} = \frac{11}{15}$

F. $\sqrt{\frac{0.0004}{0.81}} = \sqrt{\frac{4}{8100}} = \sqrt{(\frac{2}{90})^2} = \frac{2}{90} = \frac{1}{45}$

G. $\sqrt[3]{\frac{-0.001}{0.027}} = -\sqrt[3]{\frac{1}{27}} = -\frac{1}{3}$

H. $\sqrt{\frac{0.4}{0.009}} = \sqrt{\frac{400}{9}} = \sqrt{(\frac{20}{3})^2} = \frac{20}{3}$

I. $\sqrt[3]{\frac{8}{0.081}} = \sqrt[3]{\frac{8000}{81}} = \sqrt[3]{\frac{20^3}{27 \cdot 3}} = \sqrt[3]{\frac{20^3}{3^3} \cdot \frac{1}{3}} = \frac{20}{3}\sqrt[3]{\frac{1}{3}} = \frac{20}{3} \cdot \frac{1}{\sqrt[3]{3}} = \frac{20}{3} \cdot \frac{\sqrt[3]{27}}{\sqrt[3]{3} \cdot \sqrt[3]{27}} = \frac{20}{3} \cdot \frac{\sqrt[3]{27}}{3} = \frac{20\sqrt[3]{27}}{9}$

J. $\sqrt[3]{\frac{1.5}{9}} = \sqrt[3]{\frac{15}{90}} = \sqrt[3]{\frac{3}{18}} = \sqrt[3]{\frac{1}{6}} = \frac{1}{\sqrt[3]{6}} = \frac{\sqrt[3]{36}}{\sqrt[3]{6} \cdot \sqrt[3]{36}} = \frac{\sqrt[3]{36}}{6}$

Examples 5 in Radicals

Simplify each radical below or put each in a mixed radical if it can be done so. A mixed radical is a radical multiplied by a rational such as $2\sqrt{3}$ and $4\sqrt[3]{2}$.

0. $\sqrt{x^2}$

1. $\sqrt{x^3}$

2. $\sqrt{x^4}$

3. $\sqrt{x^5}$

4. $\sqrt{xy^2}$

5. $\sqrt{x^2y^2}$

6. $\sqrt{x^5y^2}$

7. $\sqrt{y^6}$

8. $\sqrt{x^5y^7}$

9. $\sqrt{|x|^2}$

A. $\sqrt{|x|^2y^2}$

B. $\sqrt[3]{x^3}$

C. $\sqrt[3]{x^4}$

D. $\sqrt[3]{x^5}$

E. $\sqrt[3]{x^7}$

F. $\sqrt[3]{-x^{23}}$

G. $\sqrt[3]{-x^3y^3}$

H. $\sqrt[3]{x^4y^4}$

I. $\sqrt[3]{-x^4y^4}$

J. $\sqrt{\dfrac{1}{x^2}}$

K. $\sqrt{\dfrac{1}{x^3}}$

L. $\sqrt{\dfrac{-2}{x^3}}$

M. $\sqrt[3]{\dfrac{8}{-x^3}}$

N. $\sqrt[3]{-\dfrac{4y^3}{x^6}}$

O. $\sqrt{\dfrac{7y^3}{8x^2}}$

P. $\sqrt{\dfrac{3x^2}{-4y}}$

Q. $\sqrt[3]{\dfrac{64y^2}{9x^3}}$

R. $\sqrt[3]{-\dfrac{16y^3z^4}{81x^4}}$

S. $\sqrt{\dfrac{4y^3z^3}{27x^3}}$

T. $\sqrt[3]{\dfrac{27z^5}{4x^4y^5}}$

76

Suggestions or Solutions
To the Examples 5 in Radicals

0. $\sqrt{x^2} = x$ if $x \geq 0$. If however, $x < 0$, we get: $\sqrt{x^2} = -x$, because $\sqrt{x^2} > 0$.

So in sum, we get: $\sqrt{x^2} = |x|$. What do we mean by $|x|$ though?

$|x|$ is the absolute value of, that is, the magnitude of x.
So for instance, we have: $|-2| = 2$, $|0| = 0$, and $|3| = 3$.

And we want to keep in mind that not only what's inside a square root sign but the square root, too, is ≥ 0.

So if for instance, $r = \sqrt{t}$, we get: $r \geq 0$, that is, $\sqrt{t} \geq 0$ as well as $t \geq 0$.

1. $\sqrt{x^3} = \sqrt{x^2 x} = x\sqrt{x}$ for $x \geq 0$.   Why not $\sqrt{x^2 x} = |x|\sqrt{x}$, though?

It is OK, too, to set: $\sqrt{x^2 x} = |x|\sqrt{x}$ for $x \geq 0$, because if $x \geq 0$, $|x|\sqrt{x} = x\sqrt{x}$.

And in $\sqrt{x^3}$, x cannot be negative, because what's inside a square root sign cannot be negative, that is, it can only be positive or 0.

So it is necessary to specify $x \geq 0$ in this case.

Just setting: $\sqrt{x^3} = \sqrt{x^2 x} = x\sqrt{x}$, we mean x can be any of all real numbers including negative numbers, that is, it can be negative as well as positive or 0.

2. $\sqrt{x^4} = \sqrt{(x^2)^2} = x^2$. Why not $|x^2|$ though?

We have: $|x^2| = x^2$, since $|x^2| \geq 0$, and so is x^2. So the absolute value sign is unnecessary.

3. $\sqrt{x^5} = \sqrt{x^4 x} = x^2 \sqrt{x}$ for $x \geq 0$.

4. $\sqrt{xy^2} = \sqrt{y^2} \sqrt{x} = |y| \sqrt{x}$ for $x \geq 0$.

5. $\sqrt{x^2 y^2} = \sqrt{x^2} \sqrt{y^2} = |x| \cdot |y| = |xy|$.

Note however, it is ***not always*** the case where $|x - y| = |x| - |y|$. So for instance, we have:

$3 - 2| = |1| = 1$, and $|3| - |2| = 3 - 2 = 1$.

$2 - 3| = |-1| = 1$, but $|2| - |3| = 2 - 3 = -1$.

So we have: $|x - y| \neq |x| - |y|$.

6. $\sqrt{x^5 y^2} = \sqrt{y^2} \sqrt{x^5} = |y| \sqrt{x^5} = |y| x^2 \sqrt{x}$ for $x \geq 0$.

7. $\sqrt{y^6} = \sqrt{(y^3)^2} = |y^3|$. That's because:

If $y \geq 0$, we get $y^3 \geq 0$. So it is the case where $\sqrt{y^6} = y^3$, because $\sqrt{y^6} \geq 0$.

If however, $y < 0$, we get: $y^3 < 0$, so we cannot just set: $\sqrt{y^6} = y^3$, because $\sqrt{y^6} > 0$.

And thus, we get: $\sqrt{y^6} = |y^3|$.

• So we want to be careful with the sign of the number or the variable when we take it out of a square root sign.

8. $\sqrt{x^5 y^7} = \sqrt{x^5}\sqrt{y^7} = \sqrt{x^4 x}\sqrt{y^6 y} = \sqrt{x^4}\sqrt{x}\sqrt{y^6}\sqrt{y} = x^2\sqrt{x}\cdot y^3\sqrt{y}$

$= x^2 y^3 \sqrt{x}\sqrt{y} = x^2 y^3 \sqrt{xy}$ for x and y both ≥ 0.

Why not $\sqrt{y^6}\sqrt{y} = |y^3|\sqrt{y}$, though?

Just setting: $\sqrt{y^6}\sqrt{y} = |y^3|\sqrt{y}$, we mean y can be any of all real numbers including negative numbers, that is, it can be negative as well as positive or 0.

And we can get the same result the way below, too:

$\sqrt{x^5 y^7} = \sqrt{x^4 y^6 xy} = \sqrt{(x^2 y^3)^2 xy} = x^2 y^3 \sqrt{xy}$ for x and y both ≥ 0.

9. $\sqrt{|x|^2}$ is nothing but $\sqrt{x^2}$, so we get: $\sqrt{|x|^2} = |x|$.

A. $\sqrt{|x|^2 y^2}$ is nothing but $\sqrt{x^2 y^2}$, so we get: $\sqrt{|x|^2 y^2} = \sqrt{x^2 y^2} = |xy|$.

B. $\sqrt[3]{x^3} = x$

C. $\sqrt[3]{x^4} = \sqrt[3]{x^3 x} = \sqrt[3]{x^3}\cdot\sqrt[3]{x} = x\sqrt[3]{x}$

D. $\sqrt[3]{x^5} = \sqrt[3]{x^3}\cdot\sqrt[3]{x^2} = x\sqrt[3]{x^2}$

E. $\sqrt[3]{x^7} = \sqrt[3]{x^6 x} = \sqrt[3]{x^6} \cdot \sqrt[3]{x} = \sqrt[3]{(x^2)^3} \cdot \sqrt[3]{x} = x^2 \sqrt[3]{x}$

F. $\sqrt[3]{-x^{23}} = -\sqrt[3]{x^{21} x^2} = -\sqrt[3]{x^{21}} \cdot \sqrt[3]{x^2} = -\sqrt[3]{(x^7)^3} \cdot \sqrt[3]{x^2} = -x^7 \sqrt[3]{x^2}$

G. $\sqrt[3]{-x^3 y^3} = -\sqrt[3]{x^3 y^3} = -\sqrt[3]{(xy)^3} = -xy$

H. $\sqrt[3]{x^4 y^4} = \sqrt[3]{x^3 y^3 xy} = \sqrt[3]{x^3 y^3} \cdot \sqrt[3]{xy} = \sqrt[3]{(xy)^3} \cdot \sqrt[3]{xy} = xy \sqrt[3]{xy}$

I. $\sqrt[3]{-x^4 y^4} = -\sqrt[3]{x^4 y^4} = -xy \sqrt[3]{xy}$

J. $\sqrt{\frac{1}{x^2}} = \frac{1}{\sqrt{x^2}} = \frac{1}{|x|} = \left| \frac{1}{x} \right|$ for $x \neq 0$.

K. $\sqrt{\frac{1}{x^3}} = \frac{1}{\sqrt{x^3}} = \frac{1}{\sqrt{x^2 x}} = \frac{1}{\sqrt{x^2}\sqrt{x}} = \frac{1}{x\sqrt{x}} = \frac{\sqrt{x}}{x\sqrt{x}\sqrt{x}} = \frac{\sqrt{x}}{x^2}$ for $x > 0$.

And we can put it the way below, too: $\sqrt{\frac{1}{x^3}} = \sqrt{\frac{x}{x^3 x}} = \frac{\sqrt{x}}{\sqrt{x^4}} = \frac{\sqrt{x}}{x^2}$ for $x > 0$.

L. $\sqrt{\frac{-2}{x^3}} = \sqrt{\frac{-2}{x} \cdot \frac{1}{x^2}} = \sqrt{\frac{-2}{x}}\sqrt{\frac{1}{x^2}} = \sqrt{\frac{-2}{x}}\sqrt{\left(\frac{1}{x}\right)^2} = -\frac{1}{x}\sqrt{\frac{-2}{x}}$ for $x < 0$. How com?

What's inside a square root sign is ≥ 0, which means in this case, x has to be < 0, and it cannot be 0, because no denominator can be 0. So x can only be negative in this case.

M. $\sqrt[3]{\frac{8}{-x^3}} = -\sqrt[3]{\left(\frac{2}{x}\right)^3} = -\frac{2}{x}$ for $x \neq 0$.

N. $\sqrt[3]{-\dfrac{4y^3}{x^6}} = -\sqrt[3]{4\cdot\dfrac{y^3}{x^6}} = -\sqrt[3]{4}\cdot\sqrt[3]{\dfrac{y^3}{x^6}} = -\sqrt[3]{4}\cdot\sqrt[3]{\left(\dfrac{y}{x^2}\right)^3} = -\sqrt[3]{4}\cdot\dfrac{y}{x^2} = -\dfrac{y\sqrt[3]{4}}{x^2}$ for $x \neq 0$.

O. $\sqrt{\dfrac{7y^3}{8x^2}} = \dfrac{\sqrt{7y^3}}{\sqrt{8x^2}} = \dfrac{\sqrt{7y^2y}}{\sqrt{4\cdot2x^2}} = \dfrac{y\sqrt{7y}}{2|x|\sqrt{2}} = \dfrac{y\sqrt{7y}\sqrt{2}}{2|x|\sqrt{2}\sqrt{2}} = \dfrac{y\sqrt{14y}}{4|x|}$ for $x \neq 0$, and $y \geq 0$.

Note that $4|x| = |4x|$.

P. $\sqrt{\dfrac{3x^2}{-4y}} = \sqrt{\dfrac{x^2}{4}}\sqrt{\dfrac{3}{-y}} = \dfrac{|x|}{2}\sqrt{\dfrac{3}{-y}} = \dfrac{|x|}{2}\sqrt{\dfrac{-3y}{yy}} = \dfrac{|x|}{2}\sqrt{\dfrac{1}{y^2}}\sqrt{-3y} = \dfrac{|x|}{-2y}\sqrt{-3y}$ for $y < 0$.

Note that in the case above, y can only be negative, because no denominator can be 0, and what's inside a square root sign is ≥ 0.

Q. $\sqrt[3]{\dfrac{64y^2}{9x^3}} = \sqrt[3]{\dfrac{4^3}{x^3}}\cdot\sqrt[3]{\dfrac{y^2}{3^2}} = \dfrac{4}{x}\sqrt[3]{\dfrac{3y^2}{3^23}} = \dfrac{4}{3x}\sqrt[3]{3y^2} = \dfrac{4\sqrt[3]{3y^2}}{3x}$ for $x \neq 0$.

R. $\sqrt[3]{-\dfrac{16y^3z^4}{81x^4}} = -\sqrt[3]{\dfrac{8\cdot2y^3z^3z}{27\cdot3x^3x}} = -\sqrt[3]{\dfrac{(2yz)^32z}{(3x)^33x}} = -\sqrt[3]{\dfrac{(2yz)^3}{(3x)^3}}\cdot\sqrt[3]{\dfrac{2z}{3x}} = -\dfrac{2yz}{3x}\sqrt[3]{\dfrac{2z}{3x}}$ for $x \neq 0$.

S. $\sqrt{\dfrac{4y^3z^3}{27x^3}} = \sqrt{\dfrac{2^2y^3z^3}{3^3x^3}} = \sqrt{\dfrac{(2yz)^2yz}{(3x)^23x}} = \dfrac{2yz}{3x}\sqrt{\dfrac{yz}{3x}} = \dfrac{2yz}{3x}\sqrt{\dfrac{yz3x}{3^2x^2}} = \dfrac{2yz}{3x}\dfrac{\sqrt{3xyz}}{3x} = \dfrac{2yz\sqrt{3xyz}}{9x^2}$ for $x > 0$, and

y and z both ≥ 0.

T. $\sqrt[3]{\dfrac{27z^5}{4x^4y^5}} = \sqrt[3]{\dfrac{2\cdot3^3z^3z^2}{8x^3y^3xy^2}} = \sqrt[3]{\dfrac{3^3z^3}{2^3x^3y^3}}\cdot\sqrt[3]{\dfrac{2z^2}{xy^2}} = \dfrac{3z}{2xy}\sqrt[3]{\dfrac{2z^2}{xy^2}} = \dfrac{3z}{2xy}\sqrt[3]{\dfrac{2yz^2}{xy^3}} = \dfrac{3z}{2xy^2}\sqrt[3]{\dfrac{2yz^2}{x}}$

$= \dfrac{3z}{2xy^2}\sqrt[3]{\dfrac{2x^2yz^2}{x^3}} = \dfrac{3z}{2x^2y^2}\sqrt[3]{2x^2yz^2} = \dfrac{3z\sqrt[3]{2x^2yz^2}}{2x^2y^2}$ for x and y both $\neq 0$.

5.0. **Problem Bases 1**

Normally, using numbers, we use real numbers, often just called real.
So unless specified otherwise, the numbers we use are real.

And using a number, we can use it in a form of a power as 2^3, which is a number, too.
So unless specified otherwise, the powers we use are real, too.

A power is though, not simply a number, but is made of two numbers called a base and an exponent. And making a power, we use real numbers as the base and the exponent.
So if we use real numbers as the base and the exponent, is the power a real number, too?
In other words, if the base and the exponent are real, is the power real, too?

It's not always the case. So it can be the case where the power is not real even if the base and the exponent both are real. It all depends on the number used as the base.
What number then, do we have to use as the base?

What matters is the sign of the number. So the sign of the base matters.
If the base is negative or 0, the power can be in trouble.   What then, is the trouble?

The power cannot be a real number. If the base is not positive, it can be the case where the power is not real. And thus, bases not positive are problematic.
What's wrong with such a base though?

The base itself is in fact, not a problem, because it cannot cause trouble alone. It depends on the exponent. If the base is not positive, we cannot use some numbers as the exponent.

• That is to say that for some exponents, we cannot use as bases numbers negative or 0.

So we want to be careful with using numbers as the exponents or the bases.
Why do we do this though?

This book is in fact, for your algebra, and is in particular, for your calculation skill, which matters when you do problems as well as when you learn things in math.
Taking care of powers or radicals when doing algebra, we do exponential algebra.
It is quite often the case, doing problems, you want to know how to work with radicals or powers, that is, how to do calculations with such numbers or expressions.
So all the material in this book is for your algebra. Knowing it very well, you can grow algebra skill fast and make it robust.

So let's take a close look at what we need to be careful with when we work with radicals or powers. We are going to cover situations where radicals or powers can be in trouble.

To begin with, if the base is 0, we cannot use a negative number as the exponent.
It's because no division by 0 is allowed. So for instance, $0^{-2} = \frac{1}{0^2} = \frac{1}{0}$ is not allowed.

And next, if the exponent is a fraction, and the denominator is even, the power can get into trouble. That is to say that when such a fractional exponent gets applied to a negative base, the power can get messed up. So a negative base is problematic, and it is particularly the case if a fractional exponent has an even denominator.

And thus, we want to be careful with situations where powers have bases negative.

Exactly what fractions then, cannot be used as the exponent if the base is negative?

Not all fractional exponents with even denominators can mess up the power.
Fractions as $\frac{3}{2}$, $\frac{1}{2}$, $-\frac{5}{4}$, $\frac{7}{4}$, -1.5, -0.5, -1.75, 0.3, 1.25, and such are not allowed to be the exponent if the base is negative.   What kind of fractions then, are they?

In such a fraction as above, the denominator is even, but the numerator is odd. How then, can the negative base mess up the power when the exponent is such a fraction?

We have in fact, already covered it in the previous section.
Let's now have one more look at though, how the problem can occur.
So to begin with, what do we mean by the problem or trouble?

The problem is that such a power troubled cannot be a real number. That is, in the real number space, which is a set of all real numbers, we do not have a power where the base is negative, and the exponent is such a fraction.   What fraction?

- It is a fraction where the denominator is even, but the numerator is odd as $\frac{1}{2}$ or $\frac{3}{4}$.

Suppose n is even, m is odd, and b is negative in a power $b^{\frac{m}{n}}$.
Then, by an exponential identity where $(x^u)^v = x^{uv}$, we can get: $b^{\frac{m}{n}} = (b^m)^{\frac{1}{n}}$. So multiplying 1 by $b^{\frac{m}{n}}$, n times, we should be able to get b^m, but we can't get it. Why not?

That's because n is even, and $b^m < 0$, since $b < 0$ and m is odd.
In the real number space, no matter what number it may be, multiplying 1 by it, even number of times, we get no negative number. We can only get a number positive or 0.

Therefore, $b^{\frac{m}{n}}$ does not exist in the real number space if $b < 0$, m is odd, and n is even.

That is, $b^{\frac{m}{n}}$ is not a real number, which is the very problem we have been talking about.

And thus, if a negative base is used, we cannot use a fractional exponent where the numerator is odd, but the denominator is even.

So let's now, take a closer look at how a power can get messed up if we apply such a fractional exponent to a base negative.

To begin with, it is *not always* the case where we can put a radical in the power form. That is to say that it is *not always* the case where we get: $\sqrt[n]{b^m} = b^{\frac{m}{n}}$. How come?

First, if $b < 0$, and m is *odd*, but n is *even*, we cannot have a power $b^{\frac{m}{n}}$.

That is to say that in the real number space, we cannot raise a number negative to a fractional exponent where the numerator is odd, but the denominator is even.

And the same is true for $\sqrt[n]{b^m}$, too, which is an nth radical, called an nth root, too.

So in the real number space, we cannot have a radical $\sqrt[n]{b^m}$ if $b < 0$, and m is odd, but n is even. That is to say that there is no real number by which we can multiply 1, even number of times to get a number negative. In other words, if in a radical, the base is negative, and the degree is even, the radical is not real.

(Note that in a radical $\sqrt[n]{b^m}$, the base is b^m, and the degree is n.)

So for instance, if $b = -1$, $m = 1$, and $n = 2$, we get: $\sqrt{-1}$, which is however, not real.

That is, we do not have a real number by which we can multiply 1 twice to get -1.

What then, do we mean by $\sqrt{-1}$?

It's a radical, of course, but is not real. And putting it in the power form, we can put it this way: $(-1)^{\frac{1}{2}}$, too. So we can do get -1 multiplying 1 by the power $(-1)^{\frac{1}{2}}$, twice or by the radical $\sqrt{-1}$, twice. Such a power or radical is however, not a real number.

It is called an *imaginary* number, and exists in a number space called the imaginary number space, which is a part of the entire number space called the complex number space.

And thus, in the imaginary number space, we have $(-1)^{\frac{1}{2}}$, which is a power not real but imaginary, which equals $\sqrt{-1}$, which is a radical not real but imaginary.

So we can put the idea above this way, too: there is no real number b for which $b^2 = \textbf{-1}$.

In fact, there is no real number b for which $b^p = c$ where $c < 0$, and p is an even integer.

Assuming for instance, $b^4 = \textbf{-27}$, we get: $b = \sqrt[4]{-27}$, which is however, not a real number.

That's because there is no real number by which we can multiply 1 four times to get -27.

Anyway, we can set: $\sqrt[4]{-27} = \sqrt[4]{(-3)^3} = (-3)^{\frac{3}{4}}$, which is not a real number though.

So in the real number space, we do not have: $\sqrt[4]{(-3)^3}$ and $(-3)^{\frac{3}{4}}$, because the base is < 0, and the exponent is a fraction where the numerator is odd, and the denominator is even.

In fact, we can't have in general, $b^{\frac{m}{n}}$ and $\sqrt[n]{b^m}$ if $b < 0$, and m is odd or irrational, but n is an even integer, in the real number space, of course.   How come?

Let's suppose $b^{\frac{m}{n}}$ is real even if $b < 0$, and m is odd or irrational, but n is even.

Then, setting $A = b^{\frac{m}{n}}$, we can say that A is real, and we get: $A = b^{\frac{m}{n}} \Rightarrow A^n = (b^{\frac{m}{n}})^n = b^m$.

So we get: $A^n = b^m$.

And we have: $b < 0$, and m is odd or irrational. So we get: $b^m < 0$, because m is not even.

However, we have: $A^n \geq 0$ no matter what real number A may be, because n is even.

Then, since we have: $A^n \geq 0$ and $b^m < 0$, we get: $A^n \neq b^m$, which however, contradicts the equality $A^n = b^m$ stated above, which in turn, contradicts the statement that A is real.

So A is not real, which means, $b^{\frac{m}{n}}$ is not real, because we have set: $A = b^{\frac{m}{n}}$.
That is to say that the supposition stated in the first place is false.

And thus, $b^{\frac{m}{n}}$ is not real if $b < 0$, m is odd or irrational, and n is even.

For instance, in the real number space, we don't have: $(-4)^{\frac{3}{2}}$, $(-1)^{\frac{1}{4}}$, $(-16)^{-\frac{1}{4}}$, and $(-9)^{-\frac{\pi}{2}}$.

Next, we have another situation where powers can get messed up when *fractional exponents* are applied to *negative bases*. So let's take a look at what the situation is.

Suppose $b < 0$, and m and n *both* are *even*. Then, we can have one of two cases below:

- Though we can have a radical $\sqrt[n]{b^m}$, we might *not* be able to have a power $b^{\frac{m}{n}}$.

- We might get: $\sqrt[n]{b^m} \neq b^{\frac{m}{n}}$ even if we can have the power $b^{\frac{m}{n}}$.

So we can say that it's *not always* the case where $A = \sqrt[n]{b^m} \Rightarrow A = b^{\frac{m}{n}}$.

We have: $A = b^{\frac{m}{n}} \Rightarrow A = \sqrt[n]{b^m}$, though.

So anyway we *cannot* have: $A = b^{\frac{m}{n}} \Leftrightarrow A = \sqrt[n]{b^m}$.

That is to say that it is *not always* the case where $\sqrt[n]{b^m} = b^{\frac{m}{n}}$.

Note that in math:

"A is true." means "A is always true."
"A is not true." means "A is untrue, or not always true."

"$X \Rightarrow Y$." means if X is true, then Y is true.
"We have $(X \Rightarrow Y)$" means "$(X \Rightarrow Y)$ is true."
"We don't have $(X \Rightarrow Y)$." means "$(X \Rightarrow Y)$ is untrue, or not always true."

"We have $(X \Leftrightarrow Y)$." means "$(X \Rightarrow Y)$ is true, and $(Y \Rightarrow X)$ is true, too."
So "$X \Leftrightarrow Y$." is the same as "$Y \Leftrightarrow X$."
"We don't have $(X \Leftrightarrow Y)$." means that either $(X \Rightarrow Y)$ is not true, or $(Y \Rightarrow X)$ is not true, or that both are not true.

5.1. **Problem Bases 2**

In the previous section, we had a situation where powers can get messed up when *fractional exponents* are applied to *negative bases*. And the situation is as follows.

Assuming $b < 0$, and m and n *both* are *even*, we can have one of two cases below:

- Though we can have a radical $\sqrt[n]{b^m}$, we might *not* be able to have a power $b^{\frac{m}{n}}$.

- We might get: $\sqrt[n]{b^m} \neq b^{\frac{m}{n}}$ even if we can have the power $b^{\frac{m}{n}}$.

Now, let's take a close look at the two cases sated above, and begin with the first case.

The first case is saying that if $b < 0$, and m and n *both* are *even*, though we can have a radical $\sqrt[n]{b^m}$, we might *not* be able to have a power $b^{\frac{m}{n}}$. That is to say that:

If $b < 0$, and m and n *both* are *even*, though a radical $\sqrt[n]{b^m}$ is real, a power $b^{\frac{m}{n}}$ *might not*.

How come?

The bottom line is, "What's inside the radical sign $\sqrt[n]{}$ has to be ≥ 0 if n is even."
How come?

First, assuming $a = \sqrt[n]{c}$, where n is even, we get: $a^n = (\sqrt[n]{c})^n = c \Rightarrow a^n = c$.

Next, we have: $a^n \geq 0$ if n is even whatever real number a may be. (e.g. $(-3)^2 = 3^2 > 0$, and $0^2 = 0$.)

So we need to have: $c \geq 0$, and c is what's inside the radical sign $\sqrt[n]{}$ where n is even.

And thus, the statement, "What's inside the radical sign $\sqrt[n]{}$ has to be ≥ 0 if n is even." can be taken as a <u>basic rule</u> for an nth radical where n is even.

So for instance, we can have:

$$\sqrt{x^2 - 2x - 3} \Rightarrow x^2 - 2x - 3 = (x - 3)(x + 1) \geq 0 \Rightarrow x \geq 3 \text{ or } x \leq -1.$$

$$\sqrt{-(x^2 - 2x - 3)} \Rightarrow x^2 - 2x - 3 = (x - 3)(x + 1) \leq 0 \Rightarrow -1 \leq x \leq 3.$$

$$\sqrt{\frac{1}{1-x}} \Rightarrow 1 - x > 0 \Rightarrow x < 1. \qquad \sqrt[4]{\frac{-1}{1-x}} \Rightarrow 1 - x < 0 \Rightarrow x > 1.$$

$$\sqrt{\frac{1}{-(x^2 - 2x - 3)}} \Rightarrow x^2 - 2x - 3 = (x - 3)(x + 1) < 0 \Rightarrow -1 < x < 3.$$

What then, about $\sqrt[3]{\dfrac{1}{1-x}}$?

"*What's inside* the radical sign $\sqrt[n]{}$ can be *any real* number if n is *odd*."

Therefore, x can be any real number except 1. Why not 1 though?

We know in the expression above, $(1 - x)$ is the denominator. So $(1 - x)$ can be any real number other than 0 since no division by 0 is allowed.

And thus, x can be any real number other than 1.

• Note that $\dfrac{1}{1-x}$ cannot be 0 for any value of x.

And let's now, practice some more algebra. This book is in fact, about algebra.
So next, for another instance, we can put the same fraction many ways as follows:

$$\sqrt[3]{\frac{1}{x^2-2x-3}} = \frac{1}{\sqrt[3]{x^2-2x-3}} = \frac{1}{(x^2-2x-3)^{\frac{1}{3}}} = (x^2-2x-3)^{-\frac{1}{3}}.$$

And we can get: $\sqrt[3]{\dfrac{1}{x^2-2x-3}} \Rightarrow x^2-2x-3 = (x-3)(x+1) \neq 0 \Rightarrow x \neq 3$ and $x \neq$ -1.

That is to say that x can be any number other than 3 and -1. How come?

We have: "*What's inside* the radical sign $\sqrt[n]{\ \ }$ can be *any real* number if n is *odd*."
So in the radical above, what's inside the radical sign can be positive, 0, or negative.

However, (x^2-2x-3) is the denominator, and thus, cannot be 0.

So (x^2-2x-3) can be any real number other than 0, (that is, positive or negative only).

Meanwhile, we have: $x^2-2x-3 = (x-3)(x+1)$.

So if $x^2-2x-3 = 0$, we get: $(x-3)(x+1) = 0 \Rightarrow x = 3$ or $x = $ -1.

That is to say that if $x^2-2x-3 \neq 0$, we get: $(x-3)(x+1) \neq 0 \Rightarrow x \neq 3$ and $x \neq$ -1.

And thus, if x^2-2x-3 can be any real number other than 0, x can be any number other than 3 and -1.

Let's now, take another example.

$$\sqrt{\frac{x}{1-x}} \Rightarrow \frac{x}{1-x} \geq 0. \text{ So we get } x(1-x) > 0 \text{ or } x = 0, \text{ but } 1-x \neq 0. \quad \text{How come?}$$

We have: "What's inside the radical sign $\sqrt[n]{}$ has to be ≥ 0 if n is even."

So we get: $\dfrac{x}{1-x} \geq 0$, because in this case we have $n = 2$, which is even.

Then, first, we can notice that $(1 - x)$ is a denominator, so we get: $1 - x \neq 0$.

Next, we can put the expression above, this way, too: $\dfrac{x}{1-x} > 0$, or $\dfrac{x}{1-x} = 0$.

And thus, taking care of $\dfrac{x}{1-x} = 0$ first, we get: $x = 0$.

Next, moving on to $\dfrac{x}{1-x} > 0$, we can notice that x and $(1 - x)$ both have the same sign.

The negative over the negative is positive, and so is the positive over the positive. And the product of two negatives is positive, and so is the product of two positives.

So we get: $x(1 - x) > 0$.

And thus, putting threads together, we have: $1 - x \neq 0$, $x = 0$, or $x(1 - x) > 0$.

That is to say that we get: $\dfrac{x}{1-x} \geq 0 \Rightarrow 1 - x \neq 0$, $x = 0$, or $x(1 - x) > 0$.

Then, beginning again with: $x(1 - x) > 0$, we get: $x(x - 1) < 0 \Rightarrow 0 < x < 1$.

Next, moving on to $1 - x \neq 0$, we get: $x \neq 1$. And we have: $x = 0$, too.

So now, putting threads together again, we have: $0 < x < 1$, $x \neq 1$, or $x = 0$.

In other words, we have: $0 \leq x < 1$.   That is to say that: $\sqrt{\dfrac{x}{1-x}} \Rightarrow 0 \leq x < 1$.

In other words, in order for $\sqrt{\dfrac{x}{1-x}}$ to be real, we have to have: $0 \leq x < 1$.

Now, getting back to the point we left off, we have the case as follows:

• If $b < 0$, and m and n *both* are *even*, though we can have a radical $\sqrt[n]{b^m}$, we might *not* be able to have a power $b^{\frac{m}{n}}$.

That is to say that the radical $\sqrt[n]{b^m}$ is real, but the power $b^{\frac{m}{n}}$ might not be real.

So let's see first, how $\sqrt[n]{b^m}$ can be real even if $b < 0$, if m and n both are even.

If n is even, and we want $\sqrt[n]{b^m}$ to exist in the real number space, we need to have: $b^m \geq 0$ due to the basic rule explained earlier. So in this case, if $b < 0$, m has to be even.

Thus, even if $b < 0$ and n is even, $\sqrt[n]{b^m}$ is valid in real number space if m, too, is even.

• That is to say that even if $b < 0$, the radical $\sqrt[n]{b^m}$ is real if m and n *both* are *even*.

For instance, we can have: $\sqrt[4]{(-7)^2} = \sqrt[4]{7^2}$, and $\sqrt[4]{(-7)^{-2}} = \sqrt[4]{\frac{1}{(-7)^2}} = \sqrt[4]{\frac{1}{7^2}}$, but *cannot* have:

$\sqrt{-6}$, $\sqrt[4]{-6}$, and $\sqrt[4]{(-2)^{-3}} = \sqrt[4]{\frac{1}{(-2)^3}} = \sqrt[4]{-\frac{1}{8}}$, in the real number space, of course.

• And of course, if n is odd, it is always the case $\sqrt[n]{b^m}$ is real for any m even if $b < 0$.

That's because we <u>can</u> get a real number by which we can multiply 1, odd number of times to get a negative number, and the real number we get is negative, too, of course.

In fact, we can have $\sqrt[n]{b^m}$ for any b nonzero and any m if n is odd, in the real number space, of course. For instance, we can have:

$$\sqrt[5]{7^{0.3}} = 7^{\frac{0.3}{5}} = 7^{\frac{3}{50}} = \sqrt[50]{7^3}, \text{ and } \sqrt[5]{(-7)^{-3}} = (-7)^{-\frac{3}{5}} = -7^{-\frac{3}{5}} = -\frac{1}{7^{\frac{3}{5}}} = -\frac{1}{\sqrt[5]{7^3}}.$$

And taking more instances, we can have:

$$\sqrt[3]{\left(-\tfrac{1}{2}\right)^{-\sqrt{5}}} = \sqrt[3]{(-2^{-1})^{\sqrt{5}}} = \sqrt[3]{-2^{-\sqrt{5}}} = (-2)^{-\frac{\sqrt{5}}{3}} = -2^{-\frac{\sqrt{5}}{3}} = -\frac{1}{2^{\frac{\sqrt{5}}{3}}} = -\frac{1}{\sqrt[3]{2^{\sqrt{5}}}}.$$

$$\sqrt[3]{-7},\ \sqrt[5]{(-7)^{-3}},\ \sqrt[3]{(-2)^9} = -2^{\frac{9}{3}} = -2^3,\ \text{and}\ \sqrt[3]{(-2)^{-9}} = \sqrt[3]{\{(-2)^{-1}\}^9} = \sqrt[3]{\left(-\tfrac{1}{2}\right)^9} = -\frac{1}{2^3} = -\tfrac{1}{8}.$$

So note that the radical is negative if the base is negative and the degree is odd.

That is, we can get this: $\sqrt[n]{c} < 0$ if $c < 0$, and n is odd.

And of course, we can get this, too: $c^{\frac{1}{n}} < 0$ if $c < 0$, and n is odd.

How come though, for instance, $\sqrt[3]{-3}$ and $\sqrt[3]{(-3)^5}$ are negative real numbers?

First, if we get a real number negative multiplying 1 by a particular number, an odd number of times, the particular number is negative and real.

For instance, we have: $1 \cdot (-2) \cdot (-2) \cdot (-2) = (-2)^3 = -8$.

So next, multiplying 1 by $\sqrt[3]{-3}$, three times, we get –3, which is negative and real.

And thus, $\sqrt[3]{-3}$ is a real number negative.

What if however, we get a negative real number multiplying 1 by a particular number, an even number of times?

Then, the particular number is not a real number, but is a number called imaginary. And imaginary numbers are available in the number space not real but imaginary. What then, do we get if we multiply 1 by an imaginary number negative, an odd number of times?

It is a number imaginary and positive. And imaginary numbers are covered in the book, **COMPLEX NUMBERS**.

And next, what do we mean by $\sqrt[n]{a}$?

We mean that multiplying 1 by $\sqrt[n]{a}$, n times, we get a.

So for instance, multiplying 1 by $\sqrt[3]{-3}$, three times, we get -3.

And thus, $\sqrt[3]{-3}$ is a negative real number, and is of course, equal to $-3^{\frac{1}{3}}$.

So next, what do we mean by $\sqrt[n]{b^m}$?

We mean that multiplying 1 by $\sqrt[n]{b^m}$, n times, we get b^m.

So for instance, multiplying 1 by $\sqrt[3]{(-3)^5}$, three times, we get $(-3)^5$, which is negative.

And thus, $\sqrt[3]{(-3)^5}$ is a negative real number, and is of course, equal to $-\sqrt[3]{3^5} = -3^{\frac{5}{3}}$.

Now, can we say that $\sqrt[4]{(-7)^2} = (-7)^{\frac{2}{4}}$?

We have: $\frac{2}{4} = \frac{1}{2}$, that is, $\sqrt[4]{(-7)^2} = \sqrt[4]{7^2} = 7^{\frac{2}{4}} = 7^{\frac{1}{2}} = \sqrt{7}$, which is real. We have this, though: $(-7)^{\frac{2}{4}} = (-7)^{\frac{1}{2}} = \sqrt{-7}$, which is not real. So we get: $\sqrt[4]{(-7)^2} \neq (-7)^{\frac{2}{4}}$.

And thus, if the base $b < 0$, $b^{\frac{m}{n}}$ might not be a real number even if m and n both are even.

That's because m is the numerator, n is the denominator in a fraction, and both are even, so simplification of the fractional exponent $\frac{m}{n}$ could end up with a fraction where the numerator is odd, but the denominator is even. For instance, $\frac{2}{6} = \frac{1}{3}$, $\frac{4}{6} = \frac{2}{3}$, but $\frac{6}{4} = \frac{3}{2}$.

So for instance, $(-3)^{\frac{2}{6}} = (-3)^{\frac{1}{3}} = \sqrt[3]{-3} = -\sqrt[3]{3} < 0$, and $(-3)^{\frac{4}{6}} = (-3)^{\frac{2}{3}} = \sqrt[3]{(-3)^2} = \sqrt[3]{9} > 0$.

However, $(-3)^{\frac{6}{4}} = (-3)^{\frac{3}{2}} = \sqrt{(-3)^3}$, which is not real. And we know: $\sqrt{(-3)^3} = \sqrt[2]{(-3)^3}$.

So we get: $\sqrt{(-3)^3} = \sqrt{-3^3} \neq -\sqrt{3^3} = -3^{\frac{3}{2}}$, simply because $-3^{\frac{3}{2}}$ is real, but $\sqrt{-3^3}$ is not. And thus, even if m and n *both* are *even*, if $b < 0$, it can be the case where $b^{\frac{m}{n}}$ is not real.

So putting threads together, in the real number space, we can say that:

If $b < 0$, and m and n *both* are *even*, though we can have a radical $\sqrt[n]{b^m}$, yet we might *not* be able to have a power $b^{\frac{m}{n}}$. So it is not always the case where $A = \sqrt[n]{b^m} \Rightarrow A = b^{\frac{m}{n}}$.

Next, let's move on to the second case as follows.

• If $b < 0$, and m and n *both* are *even*. we might get: $\sqrt[n]{b^m} \neq b^{\frac{m}{n}}$ even if we can have a power $b^{\frac{m}{n}}$, in the real number space, of course.

We show the truth of the statement above if we can come up with at least one example satisfying each of the two cases as follows: $\sqrt[n]{b^m} \neq b^{\frac{m}{n}}$, and $\sqrt[n]{b^m} = b^{\frac{m}{n}}$.

So first, can we say that $\sqrt[6]{(-8)^2} = (-8)^{\frac{2}{6}}$?

We have: $\sqrt[6]{(-8)^2} = \sqrt[6]{8^2} = 8^{\frac{2}{6}} = 8^{\frac{1}{3}} = 2$, but $(-8)^{\frac{2}{6}} = (-8)^{\frac{1}{3}} = \{(-2)^3\}^{\frac{1}{3}} = -2$.

So we get: $\sqrt[6]{(-8)^2} \neq (-8)^{\frac{2}{6}}$.

Therefore, if $b < 0$, and m and n both are even, we can get: $\sqrt[n]{b^m} \neq b^{\frac{m}{n}}$ even if $b^{\frac{m}{n}}$ is a real number if the exponent $\frac{m}{n}$ gets simplified to be a fraction where both the numerator and the denominator are odd.

Well then, can we say that $\sqrt[6]{(-8)^4} = (-8)^{\frac{4}{6}}$?

Yes, we can. We can have: $\sqrt[6]{(-8)^4} = \sqrt[6]{8^4} = 8^{\frac{4}{6}} = (2^3)^{\frac{2}{3}} = 2^2$, and also,

$(-8)^{\frac{4}{6}} = (-8)^{\frac{2}{3}} = \{(-2)^3\}^{\frac{2}{3}} = (-2)^2 = 2^2$. So we get: $\sqrt[6]{(-8)^4} = (-8)^{\frac{4}{6}}$.

So even if $b < 0$, if m and n both are even, we can get: $\sqrt[n]{b^m} = b^{\frac{m}{n}}$ if the exponent $\frac{m}{n}$ gets simplified to be a fraction where the numerator is *even*, but the denominator is *odd*.

So we can say that if $b < 0$, and m and n *both* are *even*, we might get: $\sqrt[n]{b^m} \neq b^{\frac{m}{n}}$ even if we can have a power $b^{\frac{m}{n}}$.

And thus, taking care of radicals or powers, we want to keep in mind the fact below:

If $b < 0$, and m and n *both* are *even*, we can get one of two cases below:

- Though a radical $\sqrt[n]{b^m}$ is real, the power $b^{\frac{m}{n}}$ might not be real.

- We might get: $\sqrt[n]{b^m} \neq b^{\frac{m}{n}}$ even if $b^{\frac{m}{n}}$ is real.

Now, can we say that if $m = n$, we get: $\sqrt[n]{b^m} = b$? In other words, can we get: $\sqrt[n]{b^n} = b$?

Yes, we can if $b \geq 0$.
If $b < 0$ however, we can't, yet we can still get something, which is not simple, though. What then, is it?

To begin with, if $b < 0$, and m and n both are *equal* and *even*, we get: $\sqrt[n]{b^m} = -b$.

That's because we have: $\sqrt[n]{b^m} > 0$, and $b < 0$.

More specifically, if m is *even*, and $b < 0$, we get: $b^m > 0$. So $\sqrt[n]{b^m}$ is real, and is positive.

Thus, we get: $\sqrt[n]{b^m} = -b$, because we have: $b < 0$, and $m = n \Rightarrow \frac{m}{n} = 1$.

Next, if $b < 0$, and m and n both are *equal* and *odd*, we get: $\sqrt[n]{b^m} = b$, because $\sqrt[n]{b^m} < 0$.

More specifically:

If m is *odd*, and $b < 0$, we get: $b^m < 0$, so if n is odd, too, $\sqrt[n]{b^m}$ is real, and is negative.

Thus, we get: $\sqrt[n]{b^m} = b$, because we have: $b < 0$, and $m = n \Rightarrow \frac{m}{n} = 1$.

And thus, even if $m = n$, we don't want to just assume that $\sqrt[n]{b^m} = b$, because we could get: $\sqrt[n]{b^m} = -b$. It all depends on all of m, n, and b.

So we may want to put it the way below:

If n is even, we get: $\sqrt[n]{b^n} = |b|$. If n is odd, we get: $\sqrt[n]{b^n} = b$.

For instance:

$\sqrt[4]{(-2)^4}$ is not $(-2)^{\frac{4}{4}} = (-2)^1 = -2$ but $\sqrt[4]{2^4} = 2^{\frac{4}{4}} = 2$, and thus, $\sqrt[4]{(-2)^4}$ is not -2 but 2.
And we can put it this way: |-2|.

On the other hand, we have: $\sqrt[3]{(-2)^3} = (-2)^{\frac{3}{3}} = (-2)^1 = -2$, and thus, $\sqrt[3]{(-2)^3} = -2$.

And for the same reason as above, we want to set:

- $\sqrt[n]{b^{-n}} = \left|\frac{1}{b}\right|$ if n is even, and $\sqrt[n]{b^{-n}} = \frac{1}{b}$ if n is odd.

Note:

On a radical, operations begin with what's inside the radical sign.

That is, in a radical, operations begin with the base.

And it is particularly so if such a base is a power that has a base *negative*.

So for instance, $\sqrt[2]{(-3)^6} = \sqrt[2]{3^6} = 3^{\frac{6}{2}} = 3^3$, and $\sqrt[2]{(-3)^6} \neq (-3)^{\frac{6}{2}} = (-3)^3 = -3^3 = -27$.

5.2. **Problem Bases 3**

Radical signs can be nested. Then, operations begin with what's inside the inner most radical sign. So for instance:

$$\sqrt[p]{\sqrt[q]{\sqrt[r]{\sqrt[s]{b^t}}}} = \sqrt[p]{\sqrt[q]{\sqrt[r]{\sqrt[s]{c}}}} = \sqrt[p]{\sqrt[q]{\sqrt[r]{d}}} = \sqrt[p]{\sqrt[q]{e}} = \sqrt[p]{f} = g,$$

where $c = b^t$, $d = \sqrt[s]{c}$, $e = \sqrt[r]{d}$, and $f = \sqrt[q]{e}$.

And for specific instance, $\sqrt[4]{\sqrt[3]{\sqrt[5]{\sqrt[2]{2^{120}}}}} = \sqrt[4]{\sqrt[3]{\sqrt[5]{2^{60}}}} = \sqrt[4]{\sqrt[3]{2^{12}}} = \sqrt[4]{2^4} = 2.$

What then, about $\sqrt[4]{\sqrt[2]{\sqrt[5]{\sqrt[3]{-2^{120}}}}}$?

Such a radical is not real, that is, it's not a real number. Why not though?

To begin with, we can have: $\sqrt[3]{-2^{120}} = -2^{\frac{120}{3}} = -2^{40}$, so we get: $\sqrt[4]{\sqrt[2]{\sqrt[5]{\sqrt[3]{-2^{120}}}}} = \sqrt[4]{\sqrt[2]{\sqrt[5]{-2^{40}}}}$.

Next, we can have: $\sqrt[5]{-2^{40}} = -2^{\frac{40}{5}} = -2^8$, but we cannot get: $\sqrt[4]{\sqrt[2]{-2^8}}$. Why not though?

We can't have $\sqrt[2]{-2^8}$, because the base is negative, but the degree is even.

So we can't get: $\sqrt[4]{\sqrt[2]{-2^8}}$, and in turn, we can't have: $\sqrt[4]{\sqrt[2]{\sqrt[5]{-2^{40}}}}$ and $\sqrt[4]{\sqrt[2]{\sqrt[5]{\sqrt[3]{-2^{120}}}}}$, either.

Thus, not all radicals in the form of $\sqrt[p]{\sqrt[q]{\sqrt[r]{\sqrt[s]{b^t}}}}$ where p, q, r, and s are integers are real.

In what case then, such a radical as $\sqrt[p]{\sqrt[q]{\sqrt[r]{\sqrt[s]{b^t}}}}$ is real?

Suppose $A = \sqrt[p]{\sqrt[q]{\sqrt[r]{\sqrt[s]{b^t}}}}$. Then, A is real in each of three cases below:

(If $b = 0$, and $t > 0$, A is 0 since $b^t = 0$. For the case where $t \leq 0$, refer to the section **1.0**.)

- $b > 0$.
- $b < 0$, and t is even.
- $b < 0$, and the product $pqrst$ is odd (in other words, all of p, q, r, s, and t are odd.)

Can we set: $\sqrt[p]{\sqrt[q]{\sqrt[r]{\sqrt[s]{b^t}}}} = b^{\frac{t}{pqrs}}$, though?

We can *simply* set it that way if $b > 0$. For instance, $\sqrt[4]{\sqrt[3]{\sqrt[5]{\sqrt[2]{2^{120}}}}} = 2^{\frac{120}{4 \cdot 3 \cdot 5 \cdot 2}} = 2^{\frac{120}{120}} = 2^1 = 2$. What if $b \leq 0$?

If $b = 0$, we can do so if $t > 0$. Then, we get: $0 = 0$. (For $t \leq 0$, refer to the section **1.0**.)

If $b < 0$, we can set it that way if the product $pqrst$ is odd.

That is, if $b < 0$, and all of p, q, r, s, and t are odd, we can do so. For instance:

$$\sqrt[7]{\sqrt[3]{\sqrt[5]{\sqrt[3]{(-2)^{45}}}}} = (-2)^{\frac{45}{7\cdot3\cdot5\cdot3}} = (-2)^{\frac{1}{7}} = \sqrt[7]{-2} = -\sqrt[7]{2}.$$ What if the product $pqrst$ is even?

If $b < 0$, we can set it that way if t is even, and the product $pqrs$ is odd.
That is, if $b < 0$, and all of p, q, r, and s are odd, but t is even, we can do so. For instance:

$$\sqrt[7]{\sqrt[3]{\sqrt[5]{\sqrt[3]{(-2)^{36}}}}} = (-2)^{\frac{36}{7\cdot3\cdot5\cdot3}} = 2^{\frac{4}{35}} = \sqrt[35]{2^4}.$$ What if the product $pqrs$ is even?

If $b < 0$, t is even, and the product $pqrs$ is even, we could.
Well then, what else do we need to set it that way?

In addition to the condition that $b < 0$, t is even, and $pqrs$ is even, if in the power $b^{\frac{t}{pqrs}}$, the exponent $\frac{t}{pqrs}$ gets simplified to be another fraction where the numerator is even, but the denominator is odd (e.g. $\frac{6}{9} = \frac{2}{3}$), we can.

That is, if $b < 0$, t is even, and $\dfrac{t}{pqrs}$ gets simplified to be $\dfrac{\text{even}}{\text{odd}}$, we can. For instance:

$$\sqrt[7]{\sqrt[5]{\sqrt[2]{\sqrt[3]{(-2)^{28}}}}} = (-2)^{\frac{28}{7\cdot5\cdot2\cdot3}} = (-2)^{\frac{4}{5\cdot2\cdot3}} = (-2)^{\frac{2}{15}} = 2^{\frac{2}{15}}.$$ What other case then, can we have?

We can have one more case where $\sqrt[p]{\sqrt[q]{\sqrt[r]{\sqrt[s]{b^t}}}}$ and $b^{\frac{t}{pqrs}}$ both are real, but we cannot set:

$$\sqrt[p]{\sqrt[q]{\sqrt[r]{\sqrt[s]{b^t}}}} = b^{\frac{t}{pqrs}}, \text{that is, we get: } \sqrt[p]{\sqrt[q]{\sqrt[r]{\sqrt[s]{b^t}}}} \neq b^{\frac{t}{pqrs}}.$$

What then, exactly is the case, and how come they are not equal?

In addition to the condition that $b < 0$, t is even, and $pqrs$ is even, if in the power $b^{\frac{t}{pqrs}}$, the exponent $\frac{t}{pqrs}$ gets reduced to a fraction where the numerator and denominator both are odd (e.g. $\frac{6}{10} = \frac{3}{5}$), the radical and the power both are real, but are not equal.

For instance, $\sqrt[7]{\sqrt[4]{\sqrt[5]{\sqrt[3]{(-2)^{36}}}}}$ and $(-2)^{\frac{36}{7 \cdot 4 \cdot 5 \cdot 3}}$ both are real, but are not equal. Why not?

We have: $\sqrt[7]{\sqrt[4]{\sqrt[5]{\sqrt[3]{(-2)^{36}}}}} = \sqrt[7]{\sqrt[4]{\sqrt[5]{\sqrt[3]{2^{36}}}}}$, which is positive, but we have

$(-2)^{\frac{36}{7 \cdot 4 \cdot 5 \cdot 3}} = (-2)^{\frac{9}{7 \cdot 5 \cdot 3}} = (-2)^{\frac{3}{35}}$, which is negative, so we get: $\sqrt[7]{\sqrt[4]{\sqrt[5]{\sqrt[3]{(-2)^{36}}}}} \neq (-2)^{\frac{36}{7 \cdot 4 \cdot 5 \cdot 3}}$.

Now, putting threads together, we have five cases where we can set: $\sqrt[p]{\sqrt[q]{\sqrt[r]{\sqrt[s]{b^t}}}} = b^{\frac{t}{pqrs}}$.

And they are as follows:

- If $b > 0$, we get: $\sqrt[p]{\sqrt[q]{\sqrt[r]{\sqrt[s]{b^t}}}} = b^{\frac{t}{pq \ldots rs}} > 0$.

- If $b = 0$, and $t > 0$, we get: $0 = 0$. (We can see 'why not $t \leq 0$' in the section 1.0.)

- If $b < 0$ and the product $pqrst$ is odd, we get: $\sqrt[p]{\sqrt[q]{\sqrt[r]{\sqrt[s]{b^t}}}} = b^{\frac{t}{pqrs}} < 0$.

- If $b < 0$, t is even, and the product $pqrs$ is odd, we get: $\sqrt[p]{\sqrt[q]{\sqrt[r]{\sqrt[s]{b^t}}}} = b^{\frac{t}{pqrs}} > 0$.

- If $b < 0$, t is even, and $\dfrac{t}{pqrs}$ gets reduced to $\dfrac{even}{odd}$, we get: $\sqrt[p]{\sqrt[q]{\sqrt[r]{\sqrt[s]{b^t}}}} = b^{\frac{t}{pqrs}} > 0$.

The same idea applies to powers, too. So expanding a power as $\{(b^x)^y\}^z$, the operations proceed *from inside to outside*, and it is particularly the case if the base is *negative*.

If the base is positive though, the direction of operations doesn't matter.

So for instance, we have an exponential identity where $(b^x)^y = (b^y)^x = b^{xy}$, where $b > 0$.

That is, the identity above always works if the base $b > 0$.

If the base $b < 0$ however, the identity above might not work because of the direction of the operation: the operation proceeds *from inside to outside*.

For instance, we have: $\{(-2)^4\}^{\frac{3}{2}} \neq \{(-2)^{\frac{3}{2}}\}^4$. How come?

Expanding the power $\{(-2)^{\frac{3}{2}}\}^4$, we want to process $(-2)^{\frac{3}{2}}$ first, which however, is not real, and thus, cannot even be processed. So $\{(-2)^{\frac{3}{2}}\}^4$ is not a real number.

On the other hand, we can have: $\{(-2)^4\}^{\frac{3}{4}} = (2^4)^{\frac{3}{4}} = 2^{\frac{12}{4}} = 2^3$.

So expanding $\{(-2)^4\}^{\frac{3}{4}}$, we *cannot* do this: $\{(-2)^4\}^{\frac{3}{4}} = (-2)^{\frac{12}{4}} = (-2)^3 = -8$.

Of course, we can have cases where the identity above works even if the base is negative.

We can have: $\{(-2)^3\}^{\frac{1}{3}} = \sqrt[3]{(-2)^3} = \sqrt[3]{(-2)(-2)(-2)} = \sqrt[3]{-2} \cdot \sqrt[3]{-2} \cdot \sqrt[3]{-2} = \left(\sqrt[3]{-2}\right)^3 = -2$, and

$\{(-2)^{\frac{1}{3}}\}^3 = \left(\sqrt[3]{-2}\right)^3 = -2$. Thus, we can have: $\{(-2)^3\}^{\frac{1}{3}} = \{(-2)^{\frac{1}{3}}\}^3$.

So expanding $\{(-2)^3\}^{\frac{1}{3}}$ and $\{(-2)^{\frac{1}{3}}\}^3$, we can just do it the way below:

$\{(-2)^3\}^{\frac{1}{3}} = (-2)^{3 \cdot \frac{1}{3}} = -2$, and $\{(-2)^{\frac{1}{3}}\}^3 = (-2)^{\frac{1}{3} \cdot 3} = -2$.

For another instance, we can have: $\{(-2)^{-\frac{1}{3}}\}^3 = \{(-2)^3\}^{-\frac{1}{3}}$. That's because:

We have: $(-2)^{-\frac{1}{3}} = \dfrac{1}{(-2)^{\frac{1}{3}}} = \dfrac{1}{\sqrt[3]{-2}} = \dfrac{1}{-\sqrt[3]{2}}$.

So we get: $\{(-2)^{-\frac{1}{3}}\}^3 = (\dfrac{1}{-\sqrt[3]{2}})^3 = -\dfrac{1}{(-\sqrt[3]{2})^3} = \dfrac{1}{-2} = -\dfrac{1}{2}$.

Also, we can get: $\{(-2)^3\}^{-\frac{1}{3}} = \dfrac{1}{\{(-2)^3\}^{\frac{1}{3}}} = \dfrac{1}{\sqrt[3]{(-2)^3}} = \dfrac{1}{-2} = -\dfrac{1}{2}$.

So expanding $\{(-2)^{-\frac{1}{3}}\}^3$ and $\{(-2)^3\}^{-\frac{1}{3}}$, we can just do it the way below:

$$\{(-2)^{-\frac{1}{3}}\}^3 = \{(-2)^3\}^{-\frac{1}{3}} = (-2)^{-\frac{3}{3}} = (-2)^{-1} = -\dfrac{1}{2}.$$

Now, another difference between $\sqrt[n]{b^m}$ and $b^{\frac{m}{n}}$ is that in the radical $\sqrt[n]{b^m}$, n can only be an integer ≥ 2, but in the power $b^{\frac{m}{n}}$, n can be any nonzero real number.

So for instance, we can have: $(-2)^{\frac{2}{\pi}} = \{(-2)^{\frac{1}{\pi}}\}^2 = \{(-2)^2\}^{\frac{1}{\pi}} = 4^{\frac{1}{\pi}} = 1.55468... > 0,$ where π is an irrational number, and is the circular ratio, which is $3.141592...$

However, we do not have such a radical as $\sqrt[\pi]{(-2)^2}$ because n in $\sqrt[n]{b^m}$ is an integer ≥ 2.

In particular, if $b > 0$ in $b^{\frac{m}{n}}$, n can be any nonzero real number, and m can be any real number, too. In fact, if $b > 0$ in a power b^x, x can be any real number.

So for instance, we can have: $4^{\frac{e}{\pi}} = 3.31845...,$ where e is irrational, called the Euler's number, and is $2.7182818...$ So a power b^x always works if $b > 0$.

- In other words, *bases negative* or 0 are problematic.

That is to say that if the base is negative, it can be the case where the radical or the power is not a number real but a number called imaginary.

What is an example of an imaginary number though?

The typical example is $(-1)^{\frac{1}{2}}$, which is $\sqrt{-1}$, which is often denoted by a letter *i*. Where in the world then, can $(-1)^{\frac{1}{2}}$ exist?

In the real number space, we do not have $(-1)^{\frac{1}{2}}$, which exists however, in the imaginary number space, which is a part of the complex number space, which has the real number space, too. What in the world is the complex number space, then?

The complex number space is a conceptual place where all numbers belong, and thus, is the entire number space.

For instance, we have: $(\sqrt{-1})^2 = \{(-1)^{\frac{1}{2}}\}^2 = (-1)^1 = -1$ in the complex number space.

So we have: $(\sqrt{-1})^2 = \sqrt{-1} \cdot \sqrt{-1} \neq \sqrt{(-1)(-1)} = \sqrt{(-1)^2} = \sqrt{1} = 1$.

That is, we have: $\{(-1)^{\frac{1}{2}}\}^2 \neq \{(-1)^2\}^{\frac{1}{2}}$.

For another example, we cannot have $\{(-4)^{\frac{3}{4}}\}^4$ because we don't have $(-4)^{\frac{3}{4}}$ in the real number space, so we get: $\{(-4)^{\frac{3}{4}}\}^4 \neq \{(-4)^4\}^{\frac{3}{4}} = (4^4)^{\frac{3}{4}} = 4^{\frac{12}{4}} = 4^3$.

Note that $\{(-4)^4\}^{\frac{3}{4}} \neq (-4)^{\frac{12}{4}} = (-4)^3 = -4^3$, and $\{(-4)^4\}^{\frac{3}{4}} = (4^4)^{\frac{3}{4}} = 4^{\frac{12}{4}} = 4^3$.

So we have been through quite a few cases and examples where a negative base can be a problem. And thus, let's now, put them in a summary.

Normally, using numbers, we use real numbers, which are often just called real.

And using a number, we can use it in a form of a power or a radical, which is a number, too. So unless specified otherwise, the powers or radicals we use are real, too.

Now, expressing a power with a fractional exponent, we can use either of two notations.

• In power notation, we can put it this way: $b^{\frac{m}{n}}$, which can be called the power form, too.

• In radical notation, we can put it this way: $\sqrt[n]{b^m}$, often called a radical or an nth root .

Whichever notation we may use though, we want to make sure that the power or the radical is *real*. If the base is negative or 0, the power might not be a real number. So bases not positive are problematic. What's wrong with such a base though?

First, if the base is 0, we cannot use a negative number as the exponent.

It's because no division by 0 is allowed. So for instance, $0^{-2} = 1/0^2 = 1/0$ is not allowed.

Next, if $b < 0$, and m is *odd*, but n is *even*, the power $b^{\frac{m}{n}}$ is not real.

And the same is true for a radical $\sqrt[n]{b^m}$, too.

Assuming for instance, $b^4 = -27$, we get: $b = \sqrt[4]{-27}$, which however, is not a real number.

And in general, $b^{\frac{m}{n}}$ and $\sqrt[n]{b^m}$ both are not real if $b < 0$, and m is odd or irrational, but n is an even integer.

Next, we have another situation where powers can get messed up when fractional exponents are applied to negative bases, and the situation is as follows.

Suppose $b < 0$, and m and n *both* are *even*. Then, we can have one of two cases below:

- Though a radical $\sqrt[n]{b^m}$ is real, the power $b^{\frac{m}{n}}$ might not be real.

- We might get: $\sqrt[n]{b^m} \neq b^{\frac{m}{n}}$ even if the power $b^{\frac{m}{n}}$ is real.

That is to say that we do *not always* get: $\sqrt[n]{b^m} = b^{\frac{m}{n}}$.

And the bottom line is:

"What's inside the radical sign $\sqrt[n]{}$ has to be ≥ 0 if n is even.", which can be therefore, taken as a basic rule for an nth radical where n is even. And of course:

"*What's inside* the radical sign $\sqrt[n]{}$ can be *any real* number if n is *odd*."

So to begin with, why do we get the situation below?

If the base $b < 0$, $b^{\frac{m}{n}}$ might not be a real number even if m and n both are even.

That's because m and n both are even, so simplification of the exponent $\frac{m}{n}$ could end up with a fraction where the numerator is odd, but the denominator is even.

So for instance, we get: $\sqrt[4]{(-7)^2} \neq (-7)^{\frac{2}{4}}$.

Next, what do we mean by the situation below?

If $b < 0$, and m and n *both* are *even*, though a radical $\sqrt[n]{b^m}$ is real, the power $b^{\frac{m}{n}}$ might not be real.

It means that we do not always get: $A = \sqrt[n]{b^m} \Rightarrow A = b^{\frac{m}{n}}$. In other words:

If $b < 0$, and m and n *both* are *even*. we might get: $\sqrt[n]{b^m} \neq b^{\frac{m}{n}}$ even if the power $b^{\frac{m}{n}}$ is real.

So for instance, we can get:: $\sqrt[6]{(-8)^2} \neq (-8)^{\frac{2}{6}}$.

We can have this, too, though: $\sqrt[6]{(-8)^4} = (-8)^{\frac{4}{6}} = (-8)^{\frac{2}{3}} = 8^{\frac{2}{3}}$, and $\sqrt[6]{(-8)^4} = \sqrt[6]{8^4} = \sqrt[3]{8^2}$.

So even if $b < 0$, if m and n both are even, we can get: $\sqrt[n]{b^m} = b^{\frac{m}{n}}$ if the exponent $\frac{m}{n}$ gets simplified to be a fraction where the numerator is *even* but the denominator is *odd*.

And even if $m = n$, we don't want to just assume that $\sqrt[n]{b^m} = b$, because we could get: $\sqrt[n]{b^m} = -b$. It all depends on all of m, n, and b. For instance, we have:

$$\sqrt[4]{(-2)^4} = \sqrt[4]{2^4} = 2^{\frac{4}{4}} = 2, \text{and } \sqrt[3]{-2^3} = \sqrt[3]{(-2)^3} = (-2)^{\frac{3}{3}} = (-2)^1 = -2.$$

So we may want to put it this way:

- If n is even, we get: $\sqrt[n]{b^n} = |b|$. And if n is odd, we get: $\sqrt[n]{b^n} = b$.

And for the same reason as above, we want to set:

- $\sqrt[n]{b^{-n}} = \left|\frac{1}{b}\right|$ if n is even, and $\sqrt[n]{b^{-n}} = \frac{1}{b}$ if n is odd.

On a radical, operations begin with what's inside the radical sign.
That is, in a radical, operations begin with the base.
And it is particularly so if such a base is a power that has a base *negative*.

So for instance, $\sqrt[2]{(-3)^6} = \sqrt[2]{3^6} = 3^{\frac{6}{2}} = 3^3$, and $\sqrt[2]{(-3)^6} \neq (-3)^{\frac{6}{2}} = (-3)^3 = -3^3 = -27$.

Suppose next, $A = \sqrt[p]{\sqrt[q]{\sqrt[r]{\sqrt[s]{b^t}}}}$. Then, A is real in each of three cases below:

(If $b = 0$, and $t > 0$, A is 0 since $b^t = 0$. For the case where $t \leq 0$, refer to the section **1.0**.)

- $b > 0$.
- $b < 0$, and t is even.
- $b < 0$, and the product *pqrst* is odd (in other words, all of p, q, r, s, and t are odd.)

And we have five cases where we can set: $\sqrt[p]{\sqrt[q]{\sqrt[r]{\sqrt[s]{b^t}}}} = b^{\frac{t}{pqrs}}$. And they are as follows:

- If $b > 0$, we get: $\sqrt[p]{\sqrt[q]{\sqrt[r]{\sqrt[s]{b^t}}}} = b^{\frac{t}{pq...rs}} > 0$.

- If $b = 0$, and $t > 0$, we get: $0 = 0$. (We can see 'why not $t \le 0$' in the section **1.0**.)

- If $b < 0$ and the product ***pqrst*** is odd, we get: $\sqrt[p]{\sqrt[q]{\sqrt[r]{\sqrt[s]{b^t}}}} = b^{\frac{t}{pqrs}} < 0$.

- If $b < 0$, t is even, and the product ***pqrs*** is odd, we get: $\sqrt[p]{\sqrt[q]{\sqrt[r]{\sqrt[s]{b^t}}}} = b^{\frac{t}{pqrs}} > 0$.

- If $b < 0$, t is even, and $\dfrac{t}{pqrs}$ gets reduced to $\dfrac{even}{odd}$, we get: $\sqrt[p]{\sqrt[q]{\sqrt[r]{\sqrt[s]{b^t}}}} = b^{\frac{t}{pqrs}} > 0$.

And the same idea applies to powers, too. So expanding a power as $\{(b^x)^y\}^z$, we don't want to just put it this way: b^{xyz}.

It's because if $b < 0$, it can be the case where we get: $\{(b^x)^y\}^z \ne b^{xyz}$. For instance, we cannot do this: $\{(-2)^4\}^{\frac{3}{4}} = (-2)^{\frac{12}{4}} = (-2)^3 = -8$, because we have: $\{(-2)^4\}^{\frac{3}{4}} = (2^4)^{\frac{3}{4}} = 8$.

That's because the operations proceed *from inside to outside*, and it is particularly the case if the base is *negative*. For instance, $\{(-2)^4\}^{\frac{3}{2}} \ne \{(-2)^{\frac{3}{2}}\}^4$, since $(-2)^{\frac{3}{2}}$ is not real.

If the base is positive though, the direction of operations doesn't matter.

So it seems we have quite a few to keep in mind doing exponential algebra. And doing algebra, we can hardly do much without doing exponential algebra. And doing algebra exponential, we get to use exponential identities, which are convenient tools, but are not guaranteed to work in all circumstances.

The exponential identities 1 and 2 always work if the bases are positive, but might not work if bases negative or 0 are used. That is to say that if the base is negative or 0, the power can get messed up. So bases negative or 0 are problem bases.

And it is particularly the case if the exponent is a fraction where denominator is even. So using an exponential identity, we want to make sure that the base is positive. And it is particularly the case, when we work with a power or radical where a constant, variable, or expression is used as the base or the exponent.

If the base is positive, we can use all real numbers as the exponent, and all the exponential identities work nicely, too. If the base is not positive, not all real numbers can be used as the exponent.

And thus, if the base is not a particular number but an expression, we want to check to see if the base can be negative or 0, and take care of the base accordingly. We tend to assume that bases are simply positive doing problems on powers where bases are not particular numbers but variables, constants, or expressions. So such bases are favorites with many examiners, probably including your teacher, too.

And in the next section followed by a couple of example sets on exponents and powers, we will get to see another set of powerful tools.

Such a tool is called a *logarithm*, usually called a *log*, for short, and is powered by two numbers called a base and an *antilogarithm*, called an *antilog*, for short. A log is in fact, an exponent. And the base we use making a log is no other than the base we use making a power, and the log is the exponent we use making the power. So working with logarithms, we do not want them to be in trouble. That is, we do not want logarithms to be in trouble. What do we mean by the trouble, though?

If unable to cover the set of all real numbers, logarithms are in trouble. So we want logarithms to cover the entire real number space. And thus, in logarithms, we do not want the bases to be 0 or negative.

That is, we want them to be positive only. So in logarithms, we do not let bases be 0 or negative. Besides, we don't allow the base to be 1, either. The exponent doesn't do much if the base is 1. If the base is 1, the power is 1 only no matter what the exponent may be. So if the base is 1, why bother working with logarithms? In logarithms therefore, we don't allow bases to be 1, 0, or negative. So in a logarithm, the base is always either between 0 and 1, or greater than 1.

Examples 1 on Powers

Note:

A small dot $\cdot$ is a multiplication operator unless specified otherwise,

e.g. $3 \cdot 5 = 15$.

0. Find the larger of the two numbers as follows: $\sqrt[3]{5}$ and $\sqrt[4]{10}$.

1. Put in ascending order, the three numbers as follows: $2^{\frac{1}{2}}$, $3^{\frac{1}{3}}$, and $10^{\frac{1}{4}}$.

Suggestions or Solutions

Find the larger of the two numbers as follows: $\sqrt[3]{5}$ and $\sqrt[4]{10}$.

The two numbers given are examples of nth roots, often called n^{th} radicals, too.

Usually (so not always), a radical can be converted to a power, which is composed of a base and an exponent. So we can put a radical in two notations as follows:

• In power notation, we can put it this way: $b^{\frac{m}{n}}$, which can be called the power form, too.

• In radical notation, we can put it this way: $\sqrt[n]{b^m}$, called an nth radical or an nth root .

And we can have: $A = b^{\frac{m}{n}} \Rightarrow A = \sqrt[n]{b^m}$. So we can put $b^{\frac{m}{n}}$ in $\sqrt[n]{b^m}$.

However, it's *not always* the case where $A = \sqrt[n]{b^m} \Rightarrow A = b^{\frac{m}{n}}$.

So it's *not always* true that $\sqrt[n]{b^m} = b^{\frac{m}{n}}$.

Also, a radical can be said to have two parts, which are the base and the degree.
For instance, in a radical $\sqrt[n]{A}$, n is the degree and A is the base.
So $\sqrt[3]{5}$ is of degree 3, and the base is 5.

Specifically, a radical $\sqrt[n]{b^m}$ can be called an nth radical of b^m, or n^{th} root of b^m.

Thus, we can say that a radical $\sqrt[3]{5}$ is a third radical of 5, or a third root of 5, and also, can be put in a power where the base is 5, and the exponent is $\frac{1}{3}$. So we have: $\sqrt[3]{5} = 5^{\frac{1}{3}}$. Usually though, $\sqrt[3]{5}$ is called a cube root of 5.

So what can we do about the problem in this example?

Consistency matters in math.

A radical can be said to have two parts, which are the base and the degree.
So comparing radicals, we need either of such two parts to be the same.

Now, $\sqrt[3]{5}$ is of degree 3, and $\sqrt[4]{10}$ is of degree 4.

So in this particular case, what should we make the same?

We can make degrees the same. That is, we can change their degrees so that we can get
a degree that can be common to both radicals.

Changing each of the two degrees though, we get a new base for each degree changed,
because the value of each radical needs to remain the same. And each new base will take
a form of a power.   How then, do we change the degrees?

Since the **LCM** (**L**east **C**ommon **M**ultiple) of the two degrees is $3 \cdot 4 = 12$, we may want
to set each degree equal to 12.

Then, we get: $\sqrt[3]{5} = \sqrt[12]{5^4}$, where 5^4 is the base, which takes a form of a power, and we
get: $\sqrt[4]{10} = \sqrt[12]{10^3}$, where 10^3 is the base, which takes a form of a power.

So we now have: $\sqrt[3]{5} = \sqrt[12]{5^4}$, and $\sqrt[4]{10} = \sqrt[12]{10^3}$.

Thus, the two radicals have the same degree, so we can now compare their bases.

$5^4 = (5^2)^2 = (25)(20 + 5) = 500 + 125 = 625.$

$10^3 = 1000.$

So we can see that $\sqrt[4]{10}$ is the larger.

• And we can get the same result by means of powers, too.

Setting first, each radical equal to its power equivalent, we get:

$\sqrt[3]{5} = 5^{\frac{1}{3}}$, and $\sqrt[4]{10} = 10^{\frac{1}{4}}$. So?

A power has two parts, one is a base, and the other is an exponent.
And comparing powers, we need either of the two parts to be uniform.
So in this particular case, what part do we want to make uniform?

In this case, we can make exponents uniform. That is, we can get a common exponent.
We don't actually change though, the values of the powers.

So we want to change the exponents keeping intact the values of the powers.
That is, we just modify the exponents so that the exponents share the same thing.
What then, is the same thing?

The two powers have two different exponents, $\frac{1}{3}$ and $\frac{1}{4}$, which are fractional, and have different denominators. So we want to make them have the common denominator.

The common denominator is 12, and thus, the exponents will be $\frac{4}{12}$ and $\frac{3}{12}$.

And modifying each exponent, we get a new base for each exponent modified, because the value of each power needs to remain the same. And each new base will take a form of a power.

Then, we get: $5^{\frac{1}{3}} = 5^{\frac{4}{12}} = (5^4)^{\frac{1}{12}}$, where the base takes a form of a power, which 5^4, and we get: $10^{\frac{1}{4}} = 10^{\frac{3}{12}} = (10^3)^{\frac{1}{12}}$, where the base takes a form of a power, which is 10^3.

So we now have: $5^{\frac{1}{3}} = 5^{\frac{4}{12}} = (5^4)^{\frac{1}{12}}$, and $10^{\frac{1}{4}} = 10^{\frac{3}{12}} = (10^3)^{\frac{1}{12}}$.

Now that the two powers share the same exponent, we can compare their bases.

$5^4 = (5^2)^2 = (25)(20 + 5) = 500 + 125 = 625$, and $10^3 = 1000$.

Therefore, $10^{\frac{1}{4}} = \sqrt[4]{10}$ is the larger of the two.

In short:

First, $\sqrt[3]{5} = 5^{\frac{1}{3}}$, and $\sqrt[4]{10} = 10^{\frac{1}{4}}$. Next, $5^{\frac{1}{3}} = (5^4)^{\frac{1}{12}} = 625^{\frac{1}{12}}$, and $10^{\frac{1}{4}} = (10^3)^{\frac{1}{12}} = 1000^{\frac{1}{12}}$.

Thus, $10^{\frac{1}{4}}$ is the larger.

118

Suggestions or Solutions
To the **Problem** in the Example **1**

Put in ascending order, the three numbers as follows: $2^{\frac{1}{2}}, 3^{\frac{1}{3}}$, and $10^{\frac{1}{4}}$.

Basically, this problem is no other than the one in the previous example. So this one requires just about the same work. How then, can we get this problem done?

Consistency matters in math.

The three numbers given are examples of powers, each of which is composed of two parts, a base and an exponent.
And comparing powers, we need either of the two parts to be the same.
So in this particular case, what part do we want to make the same?

In this case, we can make exponents the same. That is, we can get a common exponent. We don't actually change though, the values of the powers.

So we want to change the exponents keeping intact the values of the powers.
That is, we just modify exponents so that the exponents share the same thing.
What then, is the same thing?

The three powers given have three different exponents $\frac{1}{2}$, $\frac{1}{3}$, and $\frac{1}{4}$, which are fractions, and have different denominators. So?

So we may want to begin with the common denominator.

The common denominator is 12, and thus, the exponents will be $\frac{6}{12}$, $\frac{4}{12}$, and $\frac{3}{12}$.

So modifying the exponents keeping intact the values of the powers, we get:

$2^{\frac{1}{2}} = 2^{\frac{6}{12}}$, $3^{\frac{1}{3}} = 3^{\frac{4}{12}}$, and $10^{\frac{1}{4}} = 10^{\frac{3}{12}}$.

And modifying each exponent, we get a new base for each exponent modified, because the value of each power needs to remain the same. And each new base will take a form of a power.   How then, can we do so?

We have an exponential identity $(a^m)^n = a^{mn}$, so we can use $\frac{1}{12}$ as the common exponent.

Then, we get: $2^{\frac{6}{12}} = (2^6)^{\frac{1}{12}}$, $3^{\frac{4}{12}} = (3^4)^{\frac{1}{12}}$, and $10^{\frac{3}{12}} = (10^3)^{\frac{1}{12}}$.

So all the powers now have the same exponent, which is $\frac{1}{12}$, and have new bases.

Now that all the powers share the same exponent, we can compare their bases.

We have: $2^6 = (2^3)^2 = 8^2 = 64$, $3^4 = (3^2)^2 = 9^2 = 81$, and $10^3 = 1000$.

Therefore, $10^{\frac{1}{4}}$ is the largest, and $2^{\frac{1}{2}}$ is the smallest.

In short:

$2^{\frac{1}{2}} = (2^4)^{\frac{1}{12}} = 64^{\frac{1}{12}}$, $3^{\frac{1}{3}} = (3^4)^{\frac{1}{12}} = 81^{\frac{1}{12}}$, and $10^{\frac{1}{4}} = (10^3)^{\frac{1}{12}} = 1000^{\frac{1}{12}}$.

Therefore, $2^{\frac{1}{2}} < 3^{\frac{1}{3}} < 10^{\frac{1}{4}}$.

Examples 2 on Powers

0. Show that $\sqrt{\sqrt{2}+1} \cdot \sqrt[4]{3-2\sqrt{2}} = 1$ without squaring both sides of the equal sign.

1. Show that $\sqrt{3+\sqrt{5}} + \sqrt{2-\sqrt{3}} = \dfrac{(\sqrt{10}+\sqrt{6})}{2}$.

Suggestions or Solutions

Show that $\sqrt{\sqrt{2}+1} \cdot \sqrt[4]{3-2\sqrt{2}} = 1$ without squaring both sides of the equal sign.

$$3-2\sqrt{2} = 2-2\sqrt{2}+1 = (\sqrt{2})^2 - 2\sqrt{2}\cdot 1 + 1^2 = (\sqrt{2}-1)^2 \Rightarrow 3-2\sqrt{2} = (\sqrt{2}-1)^2.$$

So we get: $\sqrt[4]{3-2\sqrt{2}} = \sqrt[4]{(\sqrt{2}-1)^2} = (\sqrt{2}-1)^{\frac{2}{4}} = (\sqrt{2}-1)^{\frac{1}{2}} = \sqrt{\sqrt{2}-1}.$

Therefore, we get: $\sqrt{\sqrt{2}+1} \cdot \sqrt[4]{3-2\sqrt{2}} = \sqrt{\sqrt{2}+1} \cdot \sqrt{\sqrt{2}-1} = \sqrt{(\sqrt{2}+1)(\sqrt{2}-1)}$

$$= \sqrt{(\sqrt{2})^2 - 1^2} = \sqrt{2-1} = 1.$$

If not quite sure of the idea behind the processes above, follow the steps below:

Let's see first, what happens if we square both sides of the equality given.

So first, squaring both sides of the equality, we get: $(\sqrt{2}+1) \cdot \sqrt[2]{3-2\sqrt{2}} = 1.$

Squaring both sides again, we get: $(\sqrt{2}+1)^2(3-2\sqrt{2}) = 1.$

Next, expanding the left hand side, we get:

$\{(\sqrt{2})^2 + 2\sqrt{2}\cdot 1 + 1^2\}(3-2\sqrt{2}) = (3+2\sqrt{2})(3-2\sqrt{2}).$ What then?

We have a factorization identity, where $(x+y)(x-y) = x^2 - y^2.$

So we get: $(3+2\sqrt{2})(3-2\sqrt{2}) = 3^2 - (2\sqrt{2})^2 = 9-8 = 1,$ which is the right hand side of the equality given

Let's now, do this problem without squaring both sides of the equal sign.

To begin with, using the factorization identity above, we can readily get this:

$(\sqrt{2}+1)(\sqrt{2}-1) = (\sqrt{2})^2 - 1^2 = 2-1 = 1.$ So?

So it looks like we need to somehow modify $\sqrt[4]{3-2\sqrt{2}}$ into $\sqrt{\sqrt{2}-1}$.

Why $\sqrt{\sqrt{2}-1},$ though?

That's because we can get: $\sqrt{\sqrt{2}+1} \cdot \sqrt{\sqrt{2}-1} = \sqrt{(\sqrt{2}+1)(\sqrt{2}-1)} = \sqrt{1} = 1.$

So we want to convert $3-2\sqrt{2}$ into $(\sqrt{2}-1)^2.$ What do we mean by the conversion?

That is, we want to show that $3-2\sqrt{2} = (\sqrt{2}-1)^2.$ How though?

We know $(\sqrt{2}-1)^2$ is a complete square. So?

We have another factorization identity, where $x^2 - 2xy + y^2 = (x-y)^2.$ So?

So we can try putting $3-2\sqrt{2}$ in such a form of the identity above.

To begin with, we have: $\sqrt{2} = \sqrt{2} \cdot \sqrt{1}$.   So we can get: $3 - 2\sqrt{2} = 3 - 2\sqrt{2} \cdot \sqrt{1}$.

Next, taking $\sqrt{2}$ as x, and taking $\sqrt{1}$ as y in the identity where $x^2 - 2xy + y^2 = (x - y)^2$, we can expect that $3 = 2 + 1 = (\sqrt{2})^2 + (\sqrt{1})^2$.

So we get: $3 - 2\sqrt{2} = 2 - 2\sqrt{2} + 1 = (\sqrt{2})^2 - 2\sqrt{2} \cdot 1 + 1^2 = (\sqrt{2} - 1)^2$.

In short:

$3 - 2(2^{1/2}) = 2 - 2(2^{1/2}) + 1 = (2^{1/2})^2 - 2(2^{1/2})1 + 1^2 = (2^{1/2} - 1)^2 \Rightarrow 3 - 2(2^{1/2}) = (2^{1/2} - 1)^2$.

So $\{3 - 2(2^{1/2})\}^{1/4} = \{(2^{1/2} - 1)^2\}^{1/4} = (2^{1/2} - 1)^{1/2}$.

Therefore, $(2^{1/2} + 1)^{1/2}\{3 - 2(2^{1/2})\}^{1/4} = (2^{1/2} + 1)^{1/2}(2^{1/2} - 1)^{1/2} = \{(2^{1/2} + 1)(2^{1/2} - 1)\}^{1/2}$

$= \{(2^{1/2})^2 - 1^2\}^{1/2} = (2 - 1)^{1/2} = 1$.

Too hard to read?

We can put the same the way below, too:

$3 - 2\sqrt{2} = 2 - 2\sqrt{2} + 1 = (\sqrt{2})^2 - 2\sqrt{2} \cdot 1 + 1^2 = (\sqrt{2} - 1)^2 \Rightarrow 3 - 2\sqrt{2} = (\sqrt{2} - 1)^2$.

So we get: $\sqrt[4]{3 - 2\sqrt{2}} = \sqrt[4]{(\sqrt{2} - 1)^2} = (\sqrt{2} - 1)^{\frac{2}{4}} = (\sqrt{2} - 1)^{\frac{1}{2}} = \sqrt{\sqrt{2} - 1}$.

Therefore, we get: $\sqrt{\sqrt{2} + 1} \cdot \sqrt[4]{3 - 2\sqrt{2}} = \sqrt{\sqrt{2} + 1} \cdot \sqrt{\sqrt{2} - 1} = \sqrt{(\sqrt{2} + 1)(\sqrt{2} - 1)}$

$= \sqrt{(\sqrt{2})^2 - 1^2} = \sqrt{2 - 1} = 1$.

Suggestions or Solutions

To the **Problem** in the Example **1**

Show that $\sqrt{3+\sqrt{5}}+\sqrt{2-\sqrt{3}}=\dfrac{(\sqrt{10}+\sqrt{6})}{2}.$

$$3+\sqrt{5}=\frac{6+2\sqrt{5}}{2}=\frac{5+2\sqrt{5}+1}{2}=\frac{(\sqrt{5})^2+2\sqrt{5}\cdot 1+1^2}{2}=\frac{(\sqrt{5}+1)^2}{2}.$$

$$2-\sqrt{3}=\frac{4-2\sqrt{3}}{2}=\frac{3-2\sqrt{3}+1}{2}=\frac{(\sqrt{3}-1)^2}{2}.$$

$$\sqrt{3+\sqrt{5}}+\sqrt{2-\sqrt{3}}=\sqrt{\frac{(\sqrt{5}+1)^2}{2}}+\sqrt{\frac{(\sqrt{3}-1)^2}{2}}=\frac{\sqrt{5}+1}{\sqrt{2}}+\frac{\sqrt{3}-1}{\sqrt{2}}$$

$$=\frac{\sqrt{5}+1+\sqrt{3}-1}{\sqrt{2}}=\frac{\sqrt{5}+\sqrt{3}}{\sqrt{2}}=\frac{\sqrt{2}(\sqrt{5}+\sqrt{3})}{\sqrt{2}\cdot\sqrt{2}}=\frac{(\sqrt{10}+\sqrt{6})}{2}.$$

If not quite sure of the idea behind the processes above, follow the steps below:

Before we begin, we want to check some facts as follows:

It is true that $A=B\Rightarrow A^2=B^2$.   In other words, "$A=B$." means "$A^2=B^2$."

It is not always the case though, $A^2=B^2\Rightarrow A=B$.
In other words, "$A^2=B^2$." does *not always* mean "$A=B$."

If $AB\geq 0$, then we always get: $A^2=B^2\Rightarrow A=B$.
That is, if A and B both have the same sign, or both are **0**, we get: $A^2=B^2\Rightarrow A=B$.

We need to keep in mind though, "$A^2=B^2$." does not mean "$A=B$." That's because A and B can have different signs.

Now, this example doesn't keep us from taking squares of both sides of the equal sign.

So to begin with, squaring both sides of the equality given, we get:

$$(\sqrt{3+\sqrt{5}}+\sqrt{2-\sqrt{3}})^2 = 3+\sqrt{5}+2\sqrt{3+\sqrt{5}}\cdot\sqrt{2-\sqrt{3}}+2-\sqrt{3}$$

$$= 5+\sqrt{5}-\sqrt{3}+2\sqrt{3+\sqrt{5}}\cdot\sqrt{2-\sqrt{3}} = 5+\sqrt{5}-\sqrt{3}+2\sqrt{(3+\sqrt{5})(2-\sqrt{3})}.$$

Meanwhile, $(3+\sqrt{5})(2-\sqrt{3}) = 6+2\sqrt{5}-3\sqrt{3}-\sqrt{5}\cdot\sqrt{3}$

$$= 6+2\sqrt{5}-3\sqrt{3}-\sqrt{15} = 6+\sqrt{20}-\sqrt{27}-\sqrt{15}.$$

So we get: $5+\sqrt{5}-\sqrt{3}+2\sqrt{3+\sqrt{5}}\cdot\sqrt{2-\sqrt{3}} = 5+\sqrt{5}-\sqrt{3}+2(6+\sqrt{20}-\sqrt{27}-\sqrt{15})$

$= 17+\sqrt{5}-\sqrt{3}+\sqrt{40}-\sqrt{108}-\sqrt{60}$, which however, doesn't seem to equal the square of the right hand side.   Let's anyway, square the right hand side, too, though.

Then, we get: $\dfrac{(\sqrt{10}+\sqrt{6})^2}{2^2} = \dfrac{(10+2\sqrt{60}+6)}{4} = \dfrac{(16+4\sqrt{15})}{4} = 4+\sqrt{15}$, which however, doesn't even look close to the left hand side. Thus, simple squaring is not an option.
So we seem to have no choice but converting the left hand side of the given equality into the right hand side.

So let's now, take a closer look at both sides, and see what we can do.

We want to show that $\sqrt{3+\sqrt{5}}+\sqrt{2-\sqrt{3}} = \dfrac{(\sqrt{10}+\sqrt{6})}{2}$.

Looking at closely both sides, we can notice that the left hand side has $\sqrt{5}$ and $\sqrt{3}$, and so does the right hand side.   How come though?

Extracting the two radicals out of the right hand side, we get:

$$\frac{(\sqrt{10}+\sqrt{6})}{2}=\frac{(\sqrt{2}\cdot\sqrt{5}+\sqrt{2}\cdot\sqrt{3})}{2}=\frac{\sqrt{2}(\sqrt{5}+\sqrt{3})}{2}=\frac{\sqrt{2}(\sqrt{5}+\sqrt{3})}{(\sqrt{2})^2}=\frac{(\sqrt{5}+\sqrt{3})}{\sqrt{2}}.$$

Thus, we now have: $\sqrt{3+\sqrt{5}}+\sqrt{2-\sqrt{3}}=\frac{(\sqrt{5}+\sqrt{3})}{\sqrt{2}}$, which equals: $\frac{(\sqrt{10}+\sqrt{6})}{2}$, but has a bit different look.

Now, the numerator in the right hand side is the sum of two radicals $\sqrt{5}$ and $\sqrt{3}$, which shows that we need to somehow extract $\sqrt{5}$ from $\sqrt{3+\sqrt{5}}$, and extract $\sqrt{3}$ from $\sqrt{2-\sqrt{3}}$.

Then, it looks like we need to somehow remove from each of $\sqrt{3+\sqrt{5}}$ and $\sqrt{2-\sqrt{3}}$ the root sign placed outside each. How can we do that though?

Making a complete square what's inside a square root sign, we can remove the root sign.

So putting in a complete square each of $(3+\sqrt{5})$ and $(2-\sqrt{3})$, we can remove the root sign placed outside each.
However, we don't have a **2** in front of $\sqrt{5}$, and the same is true for $\sqrt{3}$, too.

That is, it is not the case where we have $2\sqrt{5}$ and $2\sqrt{3}$.

Well then, make it each have such a **2** in front of itself.   How though?

Compensation matters in math, and always works.

So multiplying each by 2, and then, dividing it by 2, we can get the result we want.

So to begin with, we can get: $3+\sqrt{5}=\frac{2(3+\sqrt{5})}{2}=\frac{6+2\sqrt{5}}{2}$.

Next, we can try putting $6+2\sqrt{5}$ in a complete square.

$$6+2\sqrt{5}=6+2\sqrt{5}\cdot1=5+2\sqrt{5}\cdot1+1=(\sqrt{5})^2+2\sqrt{5}\cdot1+1^2=(\sqrt{5}+\sqrt{1})^2.$$

Thus, we get: $3+\sqrt{5}=\dfrac{6+2\sqrt{5}}{2}=\dfrac{(\sqrt{5}+1)^2}{2}\Rightarrow 3+\sqrt{5}=\dfrac{(\sqrt{5}+1)^2}{2}=\left(\dfrac{\sqrt{5}+1}{\sqrt{2}}\right)^2.$

Let's now, move on to the other one: $2-\sqrt{3}$.

Then, to begin with, we can get: $2-\sqrt{3}=\dfrac{2(2-\sqrt{3})}{2}=\dfrac{4-2\sqrt{3}}{2}.$

Next, putting $4-2\sqrt{3}$ in a complete square, we get:

$$4-2\sqrt{3}=4-2\sqrt{3}\cdot1=3-2\sqrt{3}\cdot\sqrt{1}+1=(\sqrt{3})^2-2\sqrt{3}\cdot\sqrt{1}+(\sqrt{1})^2=(\sqrt{3}+\sqrt{1})^2.$$

Thus, we get: $2-\sqrt{3}=\dfrac{4-2\sqrt{3}}{2}=\dfrac{(\sqrt{3}-1)^2}{2}\Rightarrow 2-\sqrt{3}=\dfrac{(\sqrt{3}-1)^2}{(\sqrt{2})^2}=\left(\dfrac{\sqrt{3}-1}{\sqrt{2}}\right)^2.$

So we get: $\sqrt{3+\sqrt{5}}+\sqrt{2-\sqrt{3}}=\sqrt{\left(\dfrac{\sqrt{5}+1}{\sqrt{2}}\right)^2}+\sqrt{\left(\dfrac{\sqrt{3}-1}{\sqrt{2}}\right)^2}=\dfrac{\sqrt{5}+1}{\sqrt{2}}+\dfrac{\sqrt{3}-1}{\sqrt{2}}$

$=\dfrac{\sqrt{5}+1+\sqrt{3}-1}{\sqrt{2}}=\dfrac{\sqrt{5}+\sqrt{3}}{\sqrt{2}}.$

Thus, we get: $\sqrt{3+\sqrt{5}}+\sqrt{2-\sqrt{3}}=\dfrac{\sqrt{5}+\sqrt{3}}{\sqrt{2}}.$ And we know: $\dfrac{\sqrt{5}+\sqrt{3}}{\sqrt{2}}=\dfrac{\sqrt{10}+\sqrt{6}}{2}.$

So we get: $\sqrt{3+\sqrt{5}}+\sqrt{2-\sqrt{3}}=\dfrac{(\sqrt{10}+\sqrt{6})}{2}.$

Examples 3 on Powers

0.   Assuming that $\sqrt{x} + \dfrac{1}{\sqrt{x}} = 3$, find the values of A and B as follows.

0.0. $A = \dfrac{x^{\frac{3}{2}} + x^{-\frac{3}{2}} + 2}{x^2 + x^{-2} + 3}$

0.1. $B = x^{\frac{1}{4}} + x^{-\frac{1}{4}}$

1.   Assuming that $2^{x+2} = 3$, find the value of $\sqrt{(\frac{1}{8})^x}$.

2.   Assuming that $4^{2x} = 3 - 2\sqrt{2}$, find the value of $\dfrac{2^{5x} + 2^{-3x}}{2^x + 2^{-x}}$.

Suggestions or Solutions
To the **Problem 0** in the Example **0**

Assuming that $\sqrt{x} + \dfrac{1}{\sqrt{x}} = 3,$ **find the value of** $A = \dfrac{x^{\frac{3}{2}} + x^{-\frac{3}{2}} + 2}{x^2 + x^{-2} + 3}.$

To begin with, $x^{\frac{1}{2}} + x^{-\frac{1}{2}} = 3 \Rightarrow (x^{\frac{1}{2}} + x^{-\frac{1}{2}})^2 = x + 2 + x^{-1} = 3^2 \Rightarrow x + x^{-1} = 7$

$\Rightarrow (x + x^{-1})^2 = x^2 + 2 + x^{-2} = 7^2 \Rightarrow x^2 + x^{-2} = 47.$

Next, $x^{\frac{1}{2}} + x^{-\frac{1}{2}} = 3 \Rightarrow (x^{\frac{1}{2}} + x^{-\frac{1}{2}})^3 = x^{\frac{3}{2}} + 3(x^{\frac{1}{2}} + x^{-\frac{1}{2}}) + x^{-\frac{3}{2}} = 3^3 \Rightarrow x^{\frac{3}{2}} + x^{-\frac{3}{2}} = 18.$

Therefore, $A = \dfrac{x^{\frac{3}{2}} + x^{-\frac{3}{2}} + 2}{x^2 + x^{-2} + 3} = \dfrac{18 + 2}{47 + 3} = \dfrac{2}{5}.$

If not quite sure of the idea behind the processes above, follow the steps below:

Wouldn't it be nice if we could simply solve $\sqrt{x} + \frac{1}{\sqrt{x}} = 3$ so that we can put the value of x into the expression?

The equation given is not that easy. Even if we can get the value of x, we still need to simplify the fraction that A takes. The fraction looks asking a lot of work. So what are we going to do about this example?

Normally, problems in this kind do not expect us to solve such an equation as above. They want us to make use of though, such an equation. How?

We should be able to use in fact, the right hand side of the equation, which is the number 3, to find the value of A, of course. How then, do we make use of it?

Putting the equation in terms of powers, we get: $x^{\frac{1}{2}} + x^{-\frac{1}{2}} = 3.$

Then, we can notice that both the numerator and denominator in $A = \dfrac{x^{\frac{3}{2}} + x^{-\frac{3}{2}} + 2}{x^2 + x^{-2} + 3}$ can

be put in terms of the expression $(x^{\frac{1}{2}} + x^{-\frac{1}{2}})$, which is 3. So?

We can get the values of $(x^{\frac{3}{2}} + x^{-\frac{3}{2}})$ and $(x^2 + x^{-2})$.

Let's begin with the value of $(x^2 + x^{-2})$, simply because it looks easier than the other.

We should be able to derive the value from the equation, $x^{\frac{1}{2}} + x^{-\frac{1}{2}} = 3$, of course.

Squaring both sides of $x^{\frac{1}{2}} + x^{-\frac{1}{2}} = 3$, we get: $(x^{\frac{1}{2}} + x^{-\frac{1}{2}})^2 = x + 2 + x^{-1} = 3^2 \Rightarrow x + x^{-1} = 7$.

Then, squaring both the sides again, we get:

$(x + x^{-1})^2 = x^2 + 2 + x^{-2} = 7^2 \Rightarrow x^2 + x^{-2} = 47$, which is one of the two values we want.

Next, for the other one, we will cube both sides of $x^{\frac{1}{2}} + x^{-\frac{1}{2}} = 3$. How come?

That's because the exponents are $\frac{3}{2}$ and $-\frac{3}{2}$.

And we have a factorization identity, $(a + b)^3 = a^3 + 3ab(a + b) + b^3$.

So we get: $x^{\frac{1}{2}} + x^{-\frac{1}{2}} = 3 \Rightarrow (x^{\frac{1}{2}} + x^{-\frac{1}{2}})^3 = x^{\frac{3}{2}} + 3(x^{\frac{1}{2}} + x^{-\frac{1}{2}}) + x^{-\frac{3}{2}} = 3^3$.

And we know: $x^{\frac{1}{2}} + x^{-\frac{1}{2}} = 3$.

So we get: $x^{\frac{3}{2}} + 3(x^{\frac{1}{2}} + x^{-\frac{1}{2}}) + x^{-\frac{3}{2}} = 3^3 \Rightarrow x^{\frac{3}{2}} + 3 \cdot 3 + x^{-\frac{3}{2}} = 3^3$.

Thus, we get: $x^{\frac{3}{2}} + x^{-\frac{3}{2}} = 3^3 - 3^2 = 18$.

Therefore, $A = \dfrac{x^{\frac{3}{2}} + x^{-\frac{3}{2}} + 2}{x^2 + x^{-2} + 3} = \dfrac{18 + 2}{47 + 3} = \dfrac{20}{50} = \dfrac{2}{5}$.

Suggestions or Solutions

To the **Problem 1** in the Example **0**

Assuming that $\sqrt{x} + \dfrac{1}{\sqrt{x}} = 3,$ **find the value of** $B = x^{\frac{1}{4}} + x^{-\frac{1}{4}}.$

$$B^2 = (x^{\frac{1}{4}} + x^{-\frac{1}{4}})^2 = x^{\frac{1}{2}} + 2 + x^{-\frac{1}{2}} = 2 + 3 = 5 \Rightarrow B = \sqrt{5}, \text{ because } x^{\frac{1}{4}} + x^{-\frac{1}{4}} > 0.$$

If not quite sure of the idea behind the processes above, follow the steps below:

In this case, conversion of the equation given doesn't seem to help. So is it of no use?

If one way doesn't work, try the other way. Running math, we are in fact, finding ways, where we get to the solution, of course.

So this time, getting B squared, we can get the left hand side of $\sqrt{x} + \dfrac{1}{\sqrt{x}} = 3.$

Now, squaring B, we get: $B^2 = (x^{\frac{1}{4}} + x^{-\frac{1}{4}})^2 = x^{\frac{1}{2}} + 2 + x^{-\frac{1}{2}},$ which is $\sqrt{x} + \dfrac{1}{\sqrt{x}}.$

We know: $x^{\frac{1}{2}} + x^{-\frac{1}{2}} = 3.$ So we get: $B^2 = x^{\frac{1}{2}} + 2 + x^{-\frac{1}{2}} = 3 + 2 = 5 \Rightarrow B^2 = 5 \Rightarrow B = \pm\sqrt{5}.$

Then, since $B > 0$, we get: $B = \sqrt{5}.$ What about $-\sqrt{5}$? That is, how come: $B > 0$?

That's because both $x^{\frac{1}{4}}$ and $x^{-\frac{1}{4}} > 0$, and $B = x^{\frac{1}{4}} + x^{-\frac{1}{4}}.$ How come both $x^{\frac{1}{4}}$ and $x^{-\frac{1}{4}} > 0$?

That's becuase $x > 0.$ How come $x > 0$?

That's because if $x < 0$, both $x^{\frac{1}{4}}$ and $x^{-\frac{1}{4}}$ cannot exist in the real number space.

If $b < 0$, m is odd, and n is even, $b^{\frac{m}{n}}$ is not real. For instance, $(-2)^{\frac{1}{2}}$ is not a real number.

Besides, if $x = 0$, $x^{-\frac{1}{4}}$ cannot exist, because $x^{-\frac{1}{4}} = \dfrac{1}{x^{\frac{1}{4}}}$, and no division by 0 is allowed.

In short:

$$B^2 = (x^{\frac{1}{4}} + x^{-\frac{1}{4}})^2 = x^{\frac{1}{2}} + 2 + x^{-\frac{1}{2}} = 2 + 3 = 5 \Rightarrow B = \sqrt{5}, \text{because } x^{\frac{1}{4}} + x^{-\frac{1}{4}} > 0.$$

134

Assuming that $2^{x+2} = 3$, find the value of $\sqrt{(\frac{1}{8})^x}$.

In math, we never get anything out of nothing.
Working in math, we can get something out of something else.
Normally, doing a problem in math, we find the solution inside the problem, and not outside. Running math, we break the problem apart, and put the pieces together to form the solution.

The equation given has a power, which is 2^{x+2}, where the base is 2. So if we put in a power the radical given $\sqrt{(\frac{1}{8})^x}$, the power can probably have the same base, which is 2. So let's try such a modification.

Then, we get: $\sqrt{(\frac{1}{8})^x} = (\frac{1}{8})^{\frac{x}{2}} = 8^{-\frac{x}{2}} = (2^3)^{-\frac{x}{2}} = 2^{-\frac{3x}{2}}$.

Now, we have: $2^{x+2} = 3$, and have: $2^{-\frac{3x}{2}}$, of which we want to find the value. So?

We can notice that the two powers 2^{x+2} and $2^{-\frac{3x}{2}}$ can be put in terms of 2^x. So let's put them that way.

Then, we get: $2^{x+2} = 2^2 2^x = 4 \cdot 2^x = 3 \Rightarrow 2^x = \frac{3}{4}$, and also, $2^{-\frac{3x}{2}} = 2^{x \cdot (-\frac{3}{2})} = (2^x)^{-\frac{3}{2}}$.

Now, we can see the solution.

$2^x = \frac{3}{4} \Rightarrow (2^x)^{-\frac{3}{2}} = (\frac{3}{4})^{-\frac{3}{2}} = (\frac{4}{3})^{\frac{3}{2}} = \sqrt{(\frac{4}{3})^3}$. Therefore, we get: $\sqrt{(\frac{1}{8})^x} = \sqrt{(\frac{4}{3})^3}$.

In short:

$2^{x+2} = 2^2 2^x = 4 \cdot 2^x = 3 \Rightarrow 2^x = \frac{3}{4}$. $\sqrt{(\frac{1}{8})^x} = (\frac{1}{8})^{\frac{x}{2}} = 8^{-\frac{x}{2}} = (2^3)^{-\frac{x}{2}} = 2^{-\frac{3x}{2}} = (2^x)^{-\frac{3}{2}}$.

Thus, $\sqrt{(\frac{1}{8})^x} = (2^x)^{-\frac{3}{2}} = (\frac{3}{4})^{-\frac{3}{2}} = (\frac{4}{3})^{\frac{3}{2}} = \sqrt{(\frac{4}{3})^3}$.

Suggestions or Solutions
To the **Problem** in the Example **2**

Assuming that $4^{2x} = 3 - 2\sqrt{2}$, **find the value of** $\dfrac{2^{5x} + 2^{-3x}}{2^x + 2^{-x}}$.

$4^{2x} = 3 - 2\sqrt{2} = 3 - 2\sqrt{2} \cdot 1 = 2 - 2\sqrt{2} \cdot 1 + 1 = (\sqrt{2} - 1)^2 \Rightarrow 4^{2x} = (\sqrt{2} - 1)^2 = (2^{\frac{1}{2}} - 1)^2.$

Also, $4^{2x} = (4^x)^2 = \{(2^2)^x\}^2 = (2^{2x})^2.$

So $4^{2x} = (2^{2x})^2 = (2^{\frac{1}{2}} - 1)^2 \Rightarrow 2^{2x} = 2^{\frac{1}{2}} - 1$, since $2^{2x} > 0$ and $2^{\frac{1}{2}} > 1.$

Next, $\dfrac{2^{5x} + 2^{-3x}}{2^x + 2^{-x}} = \dfrac{2^x(2^{5x} + 2^{-3x})}{2^x(2^x + 2^{-x})} = \dfrac{2^{6x} + 2^{-2x}}{2^{2x} + 2^0} = \dfrac{(2^{2x})^3 + 2^{-2x}}{2^{2x} + 1}.$

So $\dfrac{2^{5x} + 2^{-3x}}{2^x + 2^{-x}} = \dfrac{(2^{2x})^3 + 2^{-2x}}{2^{2x} + 1}$. Besides, we have $2^{2x} = 2^{\frac{1}{2}} - 1.$

So $\dfrac{(2^{2x})^3 + 2^{-2x}}{2^{2x} + 1} = \dfrac{(2^{\frac{1}{2}} - 1)^3 + (2^{\frac{1}{2}} - 1)^{-1}}{2^{\frac{1}{2}} - 1 + 1} = \dfrac{2^{\frac{3}{2}} - 3 \cdot 2^{\frac{1}{2}}(2^{\frac{1}{2}} - 1) - 1 + (2^{\frac{1}{2}} - 1)^{-1}}{2^{\frac{1}{2}}}.$

Meanwhile, $(2^{\frac{1}{2}} - 1)^{-1} = \dfrac{1}{2^{\frac{1}{2}} - 1} = \dfrac{2^{\frac{1}{2}} + 1}{(2^{\frac{1}{2}} + 1)(2^{\frac{1}{2}} - 1)} = \dfrac{2^{\frac{1}{2}} + 1}{2 - 1} = 2^{\frac{1}{2}} + 1$, and

$2^{\frac{3}{2}} - 3 \cdot 2^{\frac{1}{2}}(2^{\frac{1}{2}} - 1) - 1 = 2 \cdot 2^{\frac{1}{2}} - 3 \cdot 2 + 3 \cdot 2^{\frac{1}{2}} - 1 = 5 \cdot 2^{\frac{1}{2}} - 7.$

Thus, $2^{\frac{3}{2}} - 3 \cdot 2^{\frac{1}{2}}(2^{\frac{1}{2}} - 1) - 1 + (2^{\frac{1}{2}} - 1)^{-1} = 5 \cdot 2^{\frac{1}{2}} - 7 + 2^{\frac{1}{2}} + 1 = 6 \cdot 2^{\frac{1}{2}} - 6.$

So we get: $\dfrac{(2^{2x})^3 + 2^{-2x}}{2^{2x} + 1} = \dfrac{2^{\frac{3}{2}} - 3 \cdot 2^{\frac{1}{2}}(2^{\frac{1}{2}} - 1) - 1 + (2^{\frac{1}{2}} - 1)^{-1}}{2^{\frac{1}{2}}} = \dfrac{6 \cdot 2^{\frac{1}{2}} - 6}{2^{\frac{1}{2}}}.$

Thus, we get: $\dfrac{2^{5x}+2^{-3x}}{2^x+2^{-x}} = \dfrac{6\cdot 2^{\frac{1}{2}}-6}{2^{\frac{1}{2}}} = \dfrac{2^{\frac{1}{2}}\cdot(6\cdot 2^{\frac{1}{2}}-6)}{2^{\frac{1}{2}}\cdot 2^{\frac{1}{2}}} = \dfrac{6\cdot 2-6\cdot 2^{\frac{1}{2}}}{2} = 3(2-2^{\frac{1}{2}}).$

If not quite sure of the idea behind the processes above, follow the steps below:

This example is not much more than the previous one. It's a bit more involved though. It just requires some more exponential algebra with powers.

Now, there is something common to both the power in the equation given and the one in the expression we want to find the value of. What then, is common?

It is a power of 2, which is 2^x. So it looks like we want to put the equation and expression in terms of 2^x, get the value of 2^x from the equation, and then, put the value into the expression. However, we may not want to do so this time.

That's because putting into the expression the value of 2^x, we will get to see the algebra seriously involved, that is, the exponential algebra will be a real big mess.
To those of you though who really love calculations much, it will be a lot of fun.

Let's see though, how it looks if we do the substitution.

First, we get: $4^{2x} = (2^2)^{2x} = 2^{4x} = 3-2\sqrt{2} \Rightarrow 2^x = (3-2\sqrt{2})^{\frac{1}{4}}.$

So next, putting it directly into the expression given, we get:

$\dfrac{(3-2\sqrt{2})^{\frac{5}{4}}+(3-2\sqrt{2})^{-\frac{3}{4}}}{(3-2\sqrt{2})^{\frac{1}{4}}+(3-2\sqrt{2})^{-\frac{1}{4}}}$, simplification of which can be an awful pain to most of us.

Of course, we can just leave it at that in the solution sheet, since it is a number, too. What if however, this problem were multiple-choice, and the list of the choices did not show the number? Besides, your teacher probably wouldn't allow much credit if you left the number as is. Whatelse then, can be done?

Running math is not for nothing. We don't like to do a lot of calculation, do we? One of good reasons for running math is that we can get around with too much of calculation.

Normally, the solution is inside the problem, so we may want to give a little closer look to the problem, particularly, the equation given, which is: $4^{2x} = 3 - 2\sqrt{2}$.

Then, we can notice that $3 - 2\sqrt{2}$ can be put in a form of a complete square. So let's break it apart, and then, put the pieces together.

$$4^{2x} = 3 - 2\sqrt{2} = 3 - 2\sqrt{2} \cdot 1 = 2 - 2\sqrt{2} \cdot 1 + 1 = (\sqrt{2} - 1)^2 \Rightarrow 4^{2x} = (\sqrt{2} - 1)^2 = (2^{\frac{1}{2}} - 1)^2.$$

Next, we want to take advantage of the result above, of course.

To begin with, let's do a little modification on 4^{2x}.

Then, we get: $4^{2x} = (4^x)^2 = \{(2^2)^x\}^2 = (2^{2x})^2$.

So we get: $4^{2x} = (2^{2x})^2 = (2^{\frac{1}{2}} - 1)^2 \Rightarrow 2^{2x} = 2^{\frac{1}{2}} - 1$. Why not $2^{2x} = \pm(2^{\frac{1}{2}} - 1)$, though?

That's because we have: $2^{2x} > 0$, and also, $2^{\frac{1}{2}} > 1$, that is, $2^{\frac{1}{2}} - 1 > 0$.

Next, let's take a closer look at the expression in question.

The expression is: $\dfrac{2^{5x} + 2^{-3x}}{2^x + 2^{-x}}$, which can be put in terms of 2^{2x}.

Thus, we may want to do some modifications on the expression above. So let's break it apart, and then, put the pieces together.

Then, we get: $\dfrac{2^{5x} + 2^{-3x}}{2^x + 2^{-x}} = \dfrac{2^x(2^{5x} + 2^{-3x})}{2^x(2^x + 2^{-x})} = \dfrac{2^{6x} + 2^{-2x}}{2^{2x} + 2^0} = \dfrac{(2^{2x})^3 + 2^{-2x}}{2^{2x} + 1}$.

So we get: $\dfrac{2^{5x} + 2^{-3x}}{2^x + 2^{-x}} = \dfrac{(2^{2x})^3 + 2^{-2x}}{2^{2x} + 1}$.

Besides, we have: $2^{2x} = 2^{\frac{1}{2}} - 1$. So let's plug it in now.

$$\frac{(2^{2x})^3 + 2^{-2x}}{2^{2x} + 1} = \frac{(2^{\frac{1}{2}} - 1)^3 + (2^{\frac{1}{2}} - 1)^{-1}}{2^{\frac{1}{2}} - 1 + 1} = \frac{2^{\frac{3}{2}} - 3 \cdot 2^{\frac{1}{2}}(2^{\frac{1}{2}} - 1) - 1 + (2^{\frac{1}{2}} - 1)^{-1}}{2^{\frac{1}{2}}}.$$ How come?

We have a factorization identity as follows:

$$(a - b)^3 = a^3 - 3a^2b + 3ab^2 - b^3 = a^3 - 3ab(a - b) - b^3.$$

Thus, putting $2^{\frac{1}{2}}$ into a, and -1 into b, we get: $(2^{\frac{1}{2}} - 1)^3 = 2^{\frac{3}{2}} - 3 \cdot 2^{\frac{1}{2}}(2^{\frac{1}{2}} - 1) - 1$.

Meanwhile, $(2^{\frac{1}{2}} - 1)^{-1} = \dfrac{1}{2^{\frac{1}{2}} - 1} = \dfrac{2^{\frac{1}{2}} + 1}{(2^{\frac{1}{2}} + 1)(2^{\frac{1}{2}} - 1)} = \dfrac{2^{\frac{1}{2}} + 1}{2 - 1} = 2^{\frac{1}{2}} + 1$, and

$$2^{\frac{3}{2}} - 3 \cdot 2^{\frac{1}{2}}(2^{\frac{1}{2}} - 1) - 1 = 2 \cdot 2^{\frac{1}{2}} - 3 \cdot 2 + 3 \cdot 2^{\frac{1}{2}} - 1 = 5 \cdot 2^{\frac{1}{2}} - 7.$$

Thus, we get: $2^{\frac{3}{2}} - 3 \cdot 2^{\frac{1}{2}}(2^{\frac{1}{2}} - 1) - 1 + (2^{\frac{1}{2}} - 1)^{-1} = 5 \cdot 2^{\frac{1}{2}} - 7 + 2^{\frac{1}{2}} + 1 = 6 \cdot 2^{\frac{1}{2}} - 6$.

So we get: $\dfrac{(2^{2x})^3 + 2^{-2x}}{2^{2x} + 1} = \dfrac{2^{\frac{3}{2}} - 3 \cdot 2^{\frac{1}{2}}(2^{\frac{1}{2}} - 1) - 1 + (2^{\frac{1}{2}} - 1)^{-1}}{2^{\frac{1}{2}}} = \dfrac{6 \cdot 2^{\frac{1}{2}} - 6}{2^{\frac{1}{2}}}$.

Therefore, we get: $\dfrac{(2^{2x})^3 + 2^{-2x}}{2^{2x} + 1} = \dfrac{6 \cdot 2^{\frac{1}{2}} - 6}{2^{\frac{1}{2}}} = \dfrac{2^{\frac{1}{2}} \cdot (6 \cdot 2^{\frac{1}{2}} - 6)}{2^{\frac{1}{2}} \cdot 2^{\frac{1}{2}}} = \dfrac{6 \cdot 2 - 6 \cdot 2^{\frac{1}{2}}}{2} = 3(2 - 2^{\frac{1}{2}})$.

Examples 4 on Powers

0. Suppose $f(x) = \dfrac{e^x - e^{-x}}{e^x + e^{-x}}$ where e is Euler's number, $f(a) = \frac{1}{2}$, and $f(b) = \frac{1}{3}$.

Then, find $f(a + b)$ where a and b are constant.

1. Suppose that x, y, and $z \neq \mathbf{0}$.

Suppose also, a and b are positive integers, $a^x = b^y = \mathbf{51}^z$, and $\frac{1}{x} + \frac{1}{y} = \frac{1}{z}$.

Then, find the value of $a + b$.

140

Suggestions or Solutions
To the **Problem** in the Example **0**

Suppose $f(x) = \dfrac{e^x - e^{-x}}{e^x + e^{-x}}$ **where e is Euler's number,** $f(a) = \frac{1}{2}$, **and** $f(b) = \frac{1}{3}$.
Then, find $f(a + b)$ **where a and b are constant.**

$$f(a+b) = \frac{e^{a+b} - e^{-(a+b)}}{e^{(a+b)} + e^{-(a+b)}}, f(a) = \frac{e^a - e^{-a}}{e^a + e^{-a}} = \frac{1}{2}, \text{ and } f(b) = \frac{e^b - e^{-b}}{e^b + e^{-b}} = \frac{1}{3}.$$

$$\frac{e^a - e^{-a}}{e^a + e^{-a}} = \frac{e^{2a} - 1}{e^{2a} + 1} = \frac{1}{2} \Rightarrow 2(e^{2a} - 1) = e^{2a} + 1 \Rightarrow e^{2a} = 3.$$

$$\frac{e^b - e^{-b}}{e^b + e^{-b}} = \frac{e^{2b} - 1}{e^{2b} + 1} = \frac{1}{3} \Rightarrow 3(e^{2b} - 1) = e^{2b} + 1 \Rightarrow e^{2b} = 2.$$

$$f(a+b) = \frac{e^{a+b} - e^{-(a+b)}}{e^{(a+b)} + e^{-(a+b)}} = \frac{e^{2(a+b)} - 1}{e^{2(a+b)} + 1} = \frac{e^{2a}e^{2b} - 1}{e^{2a}e^{2b} + 1} = \frac{2 \cdot 3 - 1}{2 \cdot 3 + 1} = \frac{5}{7}.$$

If not quite sure of the idea behind the processes above, follow the steps below:

This example is probably not easy for those of you who are not familiar with functions. If not much familiar with functions yet, refer to **BASIC FUNCTIONS** in the series of **ALGEBRA EXAMPLES**.

Now, we have: $f(x) = \dfrac{e^x - e^{-x}}{e^x + e^{-x}}$.   What number though, do we use as x?

Since no specification is given to x, x is assumed to be real.
What then, is the Euler's number e?

It is an irrational number, and is conceptually $(1 + 0)^\infty$ where 0 is called an infinitesimal which is as good as zero but not zero itself, and ∞ is called infinity. Actually:

$$e = \sum_{n=1}^{\infty} \frac{1}{n!} = 1 + \frac{1}{2} + \frac{1}{6} + \frac{1}{24} + ...,$$ which converges to 2.718181828459045..., which is an

irrational number.

It is said that about 200 years ago, Euler used his name to name the number when he was working with the number. Who is Euler?

Leonhard Paul Euler was a Swiss mathematician and physicist, but is said to had spent most of his life in Russia and Germany.

Next, what do we mean by $f(a + b)$?

It is the output for $x = a + b$.

Thus, plugging $a + b$ into the expression of the function f, we get $f(a + b)$.

In other words, replacing x in $f(x)$ with $a + b$, we get $f(a + b)$.

Specifically, we get: $f(a + b) = \dfrac{e^{a+b} - e^{-(a+b)}}{e^{(a+b)} + e^{-(a+b)}}$.

And by the same token, we get: $f(a) = \dfrac{e^a - e^{-a}}{e^a + e^{-a}} = \dfrac{1}{2}$, and $f(b) = \dfrac{e^b - e^{-b}}{e^b + e^{-b}} = \dfrac{1}{3}$. So?

Closely looking at the exponents in the expression of f, we can notice that we can put it in terms of e^{2x}.

We have: $(a^m)^n = a^{mn}$ for $a > 0$. So we get: $e^x e^x = e^{x+x} = e^{2x}$, and $e^x e^{-x} = e^{x-x} = e^0 = 1$.

So now, all we have to do is exponential algebra.

To begin with, we have: $f(x) = \dfrac{e^x - e^{-x}}{e^x + e^{-x}}$.

Multiplying both the numerator and denominator by e^x respectively, we get:

$$\frac{e^x - e^{-x}}{e^x + e^{-x}} = \frac{e^x(e^x - e^{-x})}{e^x(e^x + e^{-x})} = \frac{e^{2x} - 1}{e^{2x} + 1} \Rightarrow f(x) = \frac{e^{2x} - 1}{e^{2x} + 1}.$$

Then, we get: $f(a+b) = \dfrac{e^{2(a+b)} - 1}{e^{2(a+b)} + 1} = \dfrac{e^{2a+2b} - 1}{e^{2a+2b} + 1} = \dfrac{e^{2a}e^{2b} - 1}{e^{2a}e^{2b} + 1}.$

By the same token, we get: $\dfrac{e^a - e^{-a}}{e^a + e^{-a}} = \dfrac{e^{2a} - 1}{e^{2a} + 1} = \dfrac{1}{2} \Rightarrow 2(e^{2a} - 1) = e^{2a} + 1 \Rightarrow e^{2a} = 3$, and

$$\frac{e^b - e^{-b}}{e^b + e^{-b}} = \frac{e^{2b} - 1}{e^{2b} + 1} = \frac{1}{3} \Rightarrow 3(e^{2b} - 1) = e^{2b} + 1 \Rightarrow e^{2b} = 2.$$

Thus, we get: $f(a+b) = \dfrac{e^{2a}e^{2b} - 1}{e^{2a}e^{2b} + 1} = \dfrac{3 \cdot 2 - 1}{3 \cdot 2 + 1} = \dfrac{5}{7}.$

In short:

$$f(a+b) = \frac{e^{a+b} - e^{-(a+b)}}{e^{(a+b)} + e^{-(a+b)}}, f(a) = \frac{e^a - e^{-a}}{e^a + e^{-a}} = \frac{1}{2}, \text{ and } f(b) = \frac{e^b - e^{-b}}{e^b + e^{-b}} = \frac{1}{3}.$$

$$\frac{e^a - e^{-a}}{e^a + e^{-a}} = \frac{e^{2a} - 1}{e^{2a} + 1} = \frac{1}{2} \Rightarrow 2(e^{2a} - 1) = e^{2a} + 1 \Rightarrow e^{2a} = 3.$$

$$\frac{e^b - e^{-b}}{e^b + e^{-b}} = \frac{e^{2b} - 1}{e^{2b} + 1} = \frac{1}{3} \Rightarrow 3(e^{2b} - 1) = e^{2b} + 1 \Rightarrow e^{2b} = 2.$$

$$f(a+b) = \frac{e^{a+b} - e^{-(a+b)}}{e^{(a+b)} + e^{-(a+b)}} = \frac{e^{2(a+b)} - 1}{e^{2(a+b)} + 1} = \frac{e^{2a}e^{2b} - 1}{e^{2a}e^{2b} + 1} = \frac{2 \cdot 3 - 1}{2 \cdot 3 + 1} = \frac{5}{7}.$$

Suggestions or Solutions
To the **Problem** in the Example 1

Suppose that x, y, and $z \neq 0$.

Suppose also, a and b are positive integers, $a^x = b^y = 51^z$, and $\frac{1}{x} + \frac{1}{y} = \frac{1}{z}$.

Then, find the value of $a + b$.

$$a^x = 51^z \Rightarrow (a^x)^{\frac{1}{xz}} = (51^z)^{\frac{1}{xz}} \Rightarrow a^{\frac{x}{xz}} = 51^{\frac{z}{xz}} \Rightarrow a^{\frac{1}{z}} = 51^{\frac{1}{x}}.$$

$$b^y = 51^z \Rightarrow (b^y)^{\frac{1}{yz}} = (51^z)^{\frac{1}{yz}} \Rightarrow b^{\frac{y}{yz}} = 51^{\frac{z}{yz}} \Rightarrow b^{\frac{1}{z}} = 51^{\frac{1}{y}}.$$

Thus, we get: $a^{\frac{1}{z}} b^{\frac{1}{z}} = (ab)^{\frac{1}{z}} = 51^{\frac{1}{x}} 51^{\frac{1}{y}} = 51^{\frac{1}{x} + \frac{1}{y}} = 51^{\frac{1}{z}}$.   So, we get $ab = 51$.

By factorization of 51, we get: $51 = 3 \cdot 17$. And we know that a and b are positive integers.

Thus, $a = 3$ and $b = 17$, or $a = 17$ and $b = 3$.   Therefore, $a + b = 20$.

If not quite sure of the idea behind the processes above, follow the steps below:

Now, to begin with, we have a^x and b^y in the equation $a^x = b^y = 51^z$.

Next, the exponents x, y, and z are related to each other via the equality: $\frac{1}{x} + \frac{1}{y} = \frac{1}{z}$.

So we should be able to isolate (or extract) a and b each from the equation, and come up with an expression in terms of a and b using the equation and the equality.   How?

Modifying $a^x = b^y = 51^z$, we can set up a relation between 51 and each of a and b. How?

Actually, we have two equations in the equation, $a^x = b^y = 51^z$.

One is $a^x = 51^z$, and the other is $b^y = 51^z$.

Now, we have an exponential identity where $(u^s)^t = u^{st}$.

Therefore, we can get: $A = u^s \Rightarrow A^t = (u^s)^t = u^{st}$.

So first, using $a^x = 51^z$, we can get: $a^{\frac{1}{z}} = 51^{\frac{1}{x}}$. How?

$a^x = 51^z \Rightarrow (a^x)^{\frac{1}{xz}} = (51^z)^{\frac{1}{xz}} \Rightarrow a^{\frac{x}{xz}} = 51^{\frac{z}{xz}} \Rightarrow a^{\frac{1}{z}} = 51^{\frac{1}{x}}$.

Next, using $b^y = 51^z$, we can get: $b^{\frac{1}{z}} = 51^{\frac{1}{y}}$. How?

$b^y = 51^z \Rightarrow (b^y)^{\frac{1}{yz}} = (51^z)^{\frac{1}{yz}} \Rightarrow b^{\frac{y}{yz}} = 51^{\frac{z}{yz}} \Rightarrow b^{\frac{1}{z}} = 51^{\frac{1}{y}}$.

Now, we can see that:

• In $a^{\frac{1}{z}}$ and $b^{\frac{1}{z}}$, the exponents are the same.

• In $51^{\frac{1}{x}}$ and $51^{\frac{1}{y}}$, the bases are the same.

Taking the product of two powers with the same base, we get another power, where the base is the same base, and the exponent is the sum of the two exponents.

That is, we have an exponential identity where $u^m u^n = u^{m+n}$.

Taking a product of two powers with the same exponent, we get another power, where the base is the product of the two bases, and the exponent is the same exponent.

That is, we have another exponential identity where $u^m v^m = (uv)^m$.

Now, we have:

- In $a^{\frac{1}{z}}$ and $b^{\frac{1}{z}}$, the exponents are the same.

- In $51^{\frac{1}{x}}$ and $51^{\frac{1}{y}}$, the bases are the same.

Thus, we get: $a^{\frac{1}{z}}b^{\frac{1}{z}} = 51^{\frac{1}{x}}51^{\frac{1}{y}}$.

So we get: $(ab)^{\frac{1}{z}} = 51^{\frac{1}{x}+\frac{1}{y}} = 51^{\frac{1}{z}}$.

Therefore, we can see that $ab = 51$.

What then, about $a + b$?

We know a and b both are positive integers. So?

So factorizing the number 51, we can get all the possible integers for a and b.

We have: $51 = 3 \cdot 17$.
And we know that a and b are positive integers.

Thus, $a = 3$ and $b = 17$, or $a = 17$ and $b = 3$.

Either way, we get: $a + b = 20$.

In short:

$$a^x = 51^z \Rightarrow (a^x)^{\frac{1}{xz}} = (51^z)^{\frac{1}{xz}} \Rightarrow a^{\frac{x}{xz}} = 51^{\frac{z}{xz}} \Rightarrow a^{\frac{1}{z}} = 51^{\frac{1}{x}}.$$

$$b^y = 51^z \Rightarrow (b^y)^{\frac{1}{yz}} = (51^z)^{\frac{1}{yz}} \Rightarrow b^{\frac{y}{yz}} = 51^{\frac{z}{yz}} \Rightarrow b^{\frac{1}{z}} = 51^{\frac{1}{y}}.$$

Thus, we get: $a^{\frac{1}{z}}b^{\frac{1}{z}} = (ab)^{\frac{1}{z}} = 51^{\frac{1}{x}}51^{\frac{1}{y}} = 51^{\frac{1}{x}+\frac{1}{y}} = 51^{\frac{1}{z}}$.

So, we get $ab = 51$.

By factorization of 51, we get: $51 = 3 \cdot 17$.

And we know that a and b are positive integers.

Thus, $a = 3$ and $b = 17$, or $a = 17$ and $b = 3$.

Therefore, $a + b = 20$.

Examples 5 on Powers

This set of examples is for advanced students. And it is of course, for practice on exponential algebra, too, and will help get more strength on calculations with powers.

0. Assuming that x, a, b, and c are real, show that:
$$(1 + x^{a-b} + x^{a-c})^{-1} + (1 + x^{b-c} + x^{b-a})^{-1} + (1 + x^{c-a} + x^{c-b})^{-1} = 1.$$

1. Assuming that $a > b > c > 0$, show that $a^{2a}b^{2b}c^{2c} > a^{b+c}b^{c+a}c^{a+b}$.

2. Assuming that $b^2 = ac$, simplify the expression below:
$$a^2b^2c^2(a^3 + b^3 + c^3)^{-1}(a^{-3} + b^{-3} + c^{-3}).$$

Suggestions or Solutions
To the **Problem** in the Example **0**

Assuming that *x*, *a*, *b*, and *c* are real, show that:

$$(1 + x^{a-b} + x^{a-c})^{-1} + (1 + x^{b-c} + x^{b-a})^{-1} + (1 + x^{c-a} + x^{c-b})^{-1} = 1.$$

To begin with, setting: $x^a = p$, $x^b = q$, and $x^c = r$, we get:

$$(1 + x^{a-b} + x^{a-c})^{-1} + (1 + x^{b-c} + x^{b-a})^{-1} + (1 + x^{c-a} + x^{c-b})^{-1}$$

$$= (1 + \tfrac{p}{q} + \tfrac{p}{r})^{-1} + (1 + \tfrac{q}{r} + \tfrac{q}{p})^{-1} + (1 + \tfrac{r}{p} + \tfrac{r}{q})^{-1}.$$

Next, taking care of each term on the left hand side, we get:

$$1 + \tfrac{p}{q} + \tfrac{p}{r} = \tfrac{qr}{qr} + \tfrac{pr}{qr} + \tfrac{pq}{qr} = \tfrac{qr+pr+pq}{qr} \Rightarrow (1 + \tfrac{p}{q} + \tfrac{p}{r})^{-1} = \tfrac{qr}{qr+pr+pq}.$$

$$1 + \tfrac{q}{r} + \tfrac{q}{p} = \tfrac{pr}{pr} + \tfrac{pq}{pr} + \tfrac{qr}{pr} = \tfrac{pr+pq+qr}{pr} \Rightarrow (1 + \tfrac{q}{r} + \tfrac{q}{p})^{-1} = \tfrac{pr}{pr+pq+qr}.$$

$$1 + \tfrac{r}{p} + \tfrac{r}{q} = \tfrac{pq}{pq} + \tfrac{qr}{pq} + \tfrac{pr}{pq} = \tfrac{pq+qr+pr}{pq} \Rightarrow (1 + \tfrac{r}{p} + \tfrac{r}{q})^{-1} = \tfrac{pq}{pq+qr+pr}.$$

Thus, we get:

$$(1 + \tfrac{p}{q} + \tfrac{p}{r})^{-1} + (1 + \tfrac{q}{r} + \tfrac{q}{p})^{-1} + (1 + \tfrac{r}{p} + \tfrac{r}{q})^{-1} = \tfrac{qr}{qr+pr+pq} + \tfrac{pr}{pr+pq+qr} + \tfrac{pq}{pq+qr+pr} = \tfrac{pq+qr+rp}{pq+qr+rp} = 1.$$

If not quite sure of the idea behind the processes above, follow the steps below:

Examining first, the equality above, we can see that it is mainly made of three different powers, and the powers are x^a, x^b, and x^c. How come?

For instance, we have: $x^{a-b} = \dfrac{x^a}{x^b}$.

So other than 1 in the equality given in this problem, each of the terms can be taken as a fraction where each of the three powers above takes the position of the numerator or the denominator in alternate manner, and with the same frequency.

Thus, we can also, notice that every power is in the same situation.

So we can expect that:

The left hand side of the equality given will be eventually expressed in a fractional form. And since the right hand side is 1, what's in the numerator will be the same as what's in the denominator.

For instance, we may eventually be able to convert the left hand side of the equality into something like $\dfrac{x^{ab} + x^{bc} + x^{ca}}{x^{ab} + x^{bc} + x^{ca}}$, which equals 1.

What about the exponent -1 used in the term $(1 + x^{a-b} + x^{a-c})^{-1}$ and the other two terms?

We can just take the reciprocal of each of the three terms.

So for instance, we can set: $(1 + x^{a-b} + x^{a-c})^{-1} = \dfrac{1}{1 + x^{a-b} + x^{a-c}}$.

And using the facts above, we should be able to get 1 doing some exponential algebra on the left hand side of the quality given. So let's now, begin the algebra.

We want to show: $(1 + x^{a-b} + x^{a-c})^{-1} + (1 + x^{b-c} + x^{b-a})^{-1} + (1 + x^{c-a} + x^{c-b})^{-1} = 1$, which has quite a few powers.

Carrying around too many powers doing algebra, we are prone to errors.

So we may want to begin with simplification of those powers.

The powers used in the equality are x^a, x^b, and x^c, and are repeated often.   So?

So substituting each of those with a letter as *p*, *q*, or *r*, we can make the equality simpler and clearer, that is, more succinct so that we can do the algebra more easily.

And thus, setting: $x^a = p$, $x^b = q$, and $x^c = r$, we can simplify the left hand side, and get:

$$\left(1+\frac{p}{q}+\frac{p}{r}\right)^{-1}+\left(1+\frac{q}{r}+\frac{q}{p}\right)^{-1}+\left(1+\frac{r}{p}+\frac{r}{q}\right)^{-1}=1.$$

Next, taking care of each term on the left hand side, we get:

$$1+\frac{p}{q}+\frac{p}{r}=\frac{qr}{qr}+\frac{pr}{qr}+\frac{pq}{qr}=\frac{qr+pr+pq}{qr}\Rightarrow\left(1+\frac{p}{q}+\frac{p}{r}\right)^{-1}=\frac{qr}{qr+pr+pq}.$$

$$1+\frac{q}{r}+\frac{q}{p}=\frac{pr}{pr}+\frac{pq}{pr}+\frac{qr}{pr}=\frac{pr+pq+qr}{pr}\Rightarrow\left(1+\frac{q}{r}+\frac{q}{p}\right)^{-1}=\frac{pr}{pr+pq+qr}.$$

$$1+\frac{r}{p}+\frac{r}{q}=\frac{pq}{pq}+\frac{qr}{pq}+\frac{pr}{pq}=\frac{pq+qr+pr}{pq}\Rightarrow\left(1+\frac{r}{p}+\frac{r}{q}\right)^{-1}=\frac{pq}{pq+qr+pr}.$$

Thus, we get:

$$\left(1+\frac{p}{q}+\frac{p}{r}\right)^{-1}+\left(1+\frac{q}{r}+\frac{q}{p}\right)^{-1}+\left(1+\frac{r}{p}+\frac{r}{q}\right)^{-1}=\frac{qr}{qr+pr+pq}+\frac{pr}{pr+pq+qr}+\frac{pq}{pq+qr+pr}=\frac{pq+qr+rp}{pq+qr+rp}=1.$$

In short:

To begin with, setting: $x^a = p$, $x^b = q$, and $x^c = r$, we get:

$$\left(1 + x^{a-b} + x^{a-c}\right)^{-1} + \left(1 + x^{b-c} + x^{b-a}\right)^{-1} + \left(1 + x^{c-a} + x^{c-b}\right)^{-1}$$

$$=\left(1+\frac{p}{q}+\frac{p}{r}\right)^{-1}+\left(1+\frac{q}{r}+\frac{q}{p}\right)^{-1}+\left(1+\frac{r}{p}+\frac{r}{q}\right)^{-1}.$$

Next, taking care of each term on the left hand side, we get:

$$1+\frac{p}{q}+\frac{p}{r}=\frac{qr}{qr}+\frac{pr}{qr}+\frac{pq}{qr}=\frac{qr+pr+pq}{qr}\Rightarrow\left(1+\frac{p}{q}+\frac{p}{r}\right)^{-1}=\frac{qr}{qr+pr+pq}.$$

$$1+\frac{q}{r}+\frac{q}{p}=\frac{pr}{pr}+\frac{pq}{pr}+\frac{qr}{pr}=\frac{pr+pq+qr}{pr}\Rightarrow\left(1+\frac{q}{r}+\frac{q}{p}\right)^{-1}=\frac{pr}{pr+pq+qr}.$$

$$1+\frac{r}{p}+\frac{r}{q}=\frac{pq}{pq}+\frac{qr}{pq}+\frac{pr}{pq}=\frac{pq+qr+pr}{pq}\Rightarrow\left(1+\frac{r}{p}+\frac{r}{q}\right)^{-1}=\frac{pq}{pq+qr+pr}.$$ Thus, we get:

$$\left(1+\frac{p}{q}+\frac{p}{r}\right)^{-1}+\left(1+\frac{q}{r}+\frac{q}{p}\right)^{-1}+\left(1+\frac{r}{p}+\frac{r}{q}\right)^{-1}=\frac{qr}{qr+pr+pq}+\frac{pr}{pr+pq+qr}+\frac{pq}{pq+qr+pr}=\frac{pq+qr+rp}{pq+qr+rp}=1.$$

Suggestions or Solutions

To the **Problem** in the Example **1**

Assuming that $a > b > c > 0$, show that $a^{2a}b^{2b}c^{2c} > a^{b+c}b^{c+a}c^{a+b}$.

$a > b \Rightarrow \frac{a}{b} > 1$ and $a - b > 0$. Thus, $\left(\frac{a}{b}\right)^{a-b} > 1$.

Then, $\left(\frac{a}{b}\right)^{a-b} = \left(\frac{a}{b}\right)^{a}\left(\frac{a}{b}\right)^{-b} = a^{a}b^{-a}a^{-b}b^{b} > 1$. Thus, $a^{a}b^{b} > a^{b}b^{a}$.

$b > c \Rightarrow \frac{b}{c} > 1$ and $b - c > 0$. So $\left(\frac{b}{c}\right)^{b-c} > 1$.

Then, $\left(\frac{b}{c}\right)^{b-c} = \left(\frac{b}{c}\right)^{b}\left(\frac{b}{c}\right)^{-c} = b^{b}c^{-b}b^{-c}c^{c} > 1$. So $b^{b}c^{c} > b^{c}c^{b}$.

$a > c \Rightarrow \frac{a}{c} > 1$ and $a - c > 0$. So $\left(\frac{a}{c}\right)^{a-c} > 1$.

$\left(\frac{a}{c}\right)^{a-c} = \left(\frac{a}{c}\right)^{a}\left(\frac{a}{c}\right)^{-c} = a^{a}c^{-a}a^{-c}c^{c} > 1$. So $a^{a}c^{c} > a^{c}c^{a}$.

Now, we have: $a^{a}b^{b} > a^{b}b^{a}$, $b^{b}c^{c} > b^{c}c^{b}$, and $a^{a}c^{c} > a^{c}c^{a}$.

The product of all the terms on the left hand side is: $(a^{a}b^{b})(b^{b}c^{c})(a^{a}c^{c}) = a^{2a}b^{2b}c^{2c}$.

The product of all the terms on the right hand side is: $(a^{b}b^{a})(b^{c}c^{b})(a^{c}c^{a}) = a^{b+c}b^{a+c}c^{b+a}$.

Therefore, $a^{2a}b^{2b}c^{2c} > a^{b+c}b^{a+c}c^{b+a}$.

If not quite sure of the idea behind the processes above, follow the steps below:

This example provides another practice on exponential algebra.
In particular, the practice is on relational expressions with powers.
That is, it is about exponential inequalities.

More specifically though, the expression $a^{2a}b^{2b}c^{2c} > a^{b+c}b^{c+a}c^{a+b}$ we have to show can be called a conditional inequality. What then, is the condition?

The statement that $a > b > c > 0$ is the condition.

In this case though, the condition is not a premise as the ones we have to meet solving problems. That is, it is rather a means we can take advantage of than a burden to us.

In fact, it is the base on which the conditional inequality can hold. So we should begin with it, and also, be able to make use of it.

We want to know first though, what we need to show, and how we are going to do it.

So to begin with, examining the inequality we want to show, we can see that:

- The expression is made of some powers, and those powers have three bases, which are a, b, and c, which are all positive.

- The expression is basically composed of a^a, a^b, a^c, b^a, b^b, b^c, c^a, c^b, and c^c, so each of a, b, and c is used as a base and an exponent in a power. And also, the bases are used in the condition we want to take advantage of. So we may want to consider using each of the three as both, a base and an exponent.

- In addition, the bases are used in alternate manner, and with the same frequency. So all the bases repeat themselves in the same manner.

Next, examining the condition given, we can see it is not just an inequality, but can give us three different relational expressions as follows: $a > b > 0$, $b > c > 0$, and $a > c > 0$.

Let's now, come up with the relational expression, $a^{2a}b^{2b}c^{2c} > a^{b+c}b^{c+a}c^{a+b}$.

We can't just do so, of course. So what do we begin with?

By means of the fact that $a > b > c > 0$, we are going to make a^a, a^b, a^c, b^a, b^b, etc.

So first, we may want to begin with the inequality where $a > b > 0$.

Then, since $a > b$, we can get: $\frac{a}{b} > 1$, and $a - b > 0$. What then, can we get taking $\frac{a}{b}$ as a base, and taking $(a - b)$ as an exponent?

We can get: $(\frac{a}{b})^{a-b} > 1$.

Is the inequality above still true though, even if we have this: $0 < a - b < 1$?

For instance, we know: $3 > 2$, and $\sqrt{3} > \sqrt{2}$, so $(\frac{3}{2})^{\frac{1}{2}} = \frac{\sqrt{3}}{\sqrt{2}} > 1$. And we have: $1^{\frac{1}{4}} = \sqrt[4]{1} = 1$.

So since $\frac{a}{b} > 1$, we get: $(\frac{a}{b})^{\frac{1}{4}} > 1^{\frac{1}{4}} = 1$.

Next, we can make use of the inequality where $b > c > 0$.

Then, we can get: $\frac{b}{c} > 1$, and $b - c > 0$.

What then, can we get taking $\frac{b}{c}$ as a base, and taking $(b - c)$ as an exponent?

We can get: $(\frac{b}{c})^{b-c} > 1$.

And next, we have: $a > c > 0$.   So we get: $\frac{a}{c} > 1$, and $a - c > 0$.

What then, can we get taking $\frac{a}{c}$ as a base, and taking $(a - c)$ as an exponent?

We can get: $(\frac{a}{c})^{a-c} > 1$.

So we now have: $\left(\frac{a}{b}\right)^{a-b} > 1$, $\left(\frac{b}{c}\right)^{b-c} > 1$, and $\left(\frac{a}{c}\right)^{a-c} > 1$.

Next, we are going to refine each of the expressions above.

To begin with, we have:

$\left(\frac{a}{b}\right)^{a-b} = \left(\frac{a}{b}\right)^{a}\left(\frac{a}{b}\right)^{-b} = (a^1 b^{-1})^a (a^1 b^{-1})^{-b} = a^a b^{-a} a^{-b} b^b = a^{a-b} b^{-a+b} > 1$. Thus:

Multiplying both sides by $a^b b^a$ respectively, we get: $a^{a-b+b} b^{-a+b+a} > a^b b^a \Rightarrow a^a b^b > a^b b^a$.

Next, we have:

$\left(\frac{b}{c}\right)^{b-c} = \left(\frac{b}{c}\right)^{b}\left(\frac{b}{c}\right)^{-c} = (b^1 c^{-1})^b (b^1 c^{-1})^{-c} = b^b c^{-b} b^{-c} c^c = b^{b-c} c^{-b+c} > 1$. Thus:

Multiplying both sides by $b^c c^b$ respectively, we get: $b^{b-c+c} c^{-b+c+b} > b^c c^b \Rightarrow b^b c^c > b^c c^b$.

Next, we have:

$\left(\frac{a}{c}\right)^{a-c} = \left(\frac{a}{c}\right)^{a}\left(\frac{a}{c}\right)^{-c} = (a^1 c^{-1})^a (a^1 c^{-1})^{-c} = a^a c^{-a} a^{-c} c^c = a^{a-c} c^{-a+c} > 1$. Thus:

Multiplying both sides by $a^c c^a$ respectively, we get: $a^{a-c+c} c^{-a+c+a} > a^c c^a \Rightarrow a^a c^c > a^c c^a$.

Now, we have: $a^a b^b > a^b b^a$, $b^b c^c > b^c c^b$, and $a^a c^c > a^c c^a$.

So taking the product of all the terms on the left hand side, we get $a^a b^b b^b c^c a^a c^c$, which equals $a^{2a} b^{2b} c^{2c}$.

Next, taking the product of all the terms on the right hand side, we get $a^b b^a b^c c^b a^c c^a$, which equals $a^{b+c} b^{a+c} c^{b+a}$.

Therefore, we get: $a^{2a} b^{2b} c^{2c} > a^{b+c} b^{a+c} c^{b+a}$.

Suggestions or Solutions
To the **Problem** in the Example **2**

Assuming that $b^2 = ac$, simplify the expression below:
$a^2b^2c^2(a^3 + b^3 + c^3)^{-1}(a^{-3} + b^{-3} + c^{-3}).$

We have: $a^2b^2c^2(a^3 + b^3 + c^3)^{-1}(a^{-3} + b^{-3} + c^{-3}) = \dfrac{a^2b^2c^2(a^{-3} + b^{-3} + c^{-3})}{a^3 + b^3 + c^3}.$

Meanwhile, $a^2b^2c^2(a^{-3} + b^{-3} + c^{-3}) = a^{-1}b^2c^2 + a^2b^{-1}c^2 + a^2b^2c^{-1}.$

So we get:

$b^2 = ac \Rightarrow a^{-1}b^2c^2 = a^{-1}(ac)c^2 = c^3.$

$b^2 = ac \Rightarrow b^4 = a^2c^2 \Rightarrow a^2b^{-1}c^2 = a^2c^2b^{-1} = b^4b^{-1} = b^3.$

$b^2 = ac \Rightarrow a = b^2c^{-1} \Rightarrow a^2b^2c^{-1} = a^2a = a^3.$

Thus, $a^2b^2c^2(a^3 + b^3 + c^3)^{-1}(a^{-3} + b^{-3} + c^{-3}) = a^2b^2c^2(a^{-3} + b^{-3} + c^{-3})(a^3 + b^3 + c^3)^{-1}$

$= (c^3 + b^3 + a^3)(a^3 + b^3 + c^3)^{-1} = 1.$

If not quite sure of the idea behind the processes above, follow the steps below:

The expression to be simplified is the same as $a^2b^2c^2(a^{-3} + b^{-3} + c^{-3})/(a^3 + b^3 + c^3)$, which

equals $\dfrac{a^2b^2c^2(a^{-3} + b^{-3} + c^{-3})}{a^3 + b^3 + c^3}$, which however, does not look like we can simplify any

further. It just seems to be no more than a fractional expression if we just look at it.

Examining the expression closely though, we can see that the constants *a*, *b*, and *c* are used with the same frequency, and exactly in the same manner. So we can expect some similarity between the constants.

Besides, we have a condition on the expression, which provides us with a connective expression among the constants. So we should take advantage of it. Taking the benefit of it though, we may want to begin with expanding the numerator in the expression below:

$$\frac{a^2 b^2 c^2 (a^{-3} + b^{-3} + c^{-3})}{a^3 + b^3 + c^3}$$

Then, we get: $a^2 b^2 c^2 (a^{-3} + b^{-3} + c^{-3}) = a^{-1} b^2 c^2 + a^2 b^{-1} c^2 + a^2 b^2 c^{-1}$. What then, is the next?

We can now use the condition: $b^2 = ac$.

We may not just want to take though, the condition as is.
But we may want to take it in many different ways, too.
In other words, we should be able to convert the condition as many ways as necessary.

For instance, we can convert it into $c = b^2 a^{-1}$, $a = b^2 c^{-1}$, $b^4 = a^2 c^2$, etc.

Let's now, apply the condition to the numerator of $\dfrac{a^2 b^2 c^2 (a^{-3} + b^{-3} + c^{-3})}{a^3 + b^3 + c^3}$.

To begin with, we have $a^{-1} b^2 c^2$ term in the numerator.

So we can get: $b^2 = ac \Rightarrow a^{-1} b^2 c^2 = a^{-1}(ac)c^2 = c^3$.

Next, we have $a^2 b^{-1} c^2$ term.
And also, we can have: $b^2 = ac \Rightarrow b^4 = a^2 c^2$.

Thus, we can get: $b^4 = a^2 c^2 \Rightarrow a^2 b^{-1} c^2 = a^2 c^2 b^{-1} = b^4 b^{-1} = b^3$.

Next, we have $a^2 b^2 c^{-1}$ term, and also, can have: $b^2 = ac \Rightarrow a = b^2 c^{-1}$.

Thus, we can get: $a = b^2 c^{-1} \Rightarrow a^2 b^2 c^{-1} = a^2 a = a^3$.

So we can see that $a^2b^2c^2(a^{-3} + b^{-3} + c^{-3}) = a^{-1}b^2c^2 + a^2b^{-1}c^2 + a^2b^2c^{-1} = a^3 + b^3 + c^3$.

Thus, we can get: $\dfrac{a^2b^2c^2(a^{-3} + b^{-3} + c^{-3})}{a^3 + b^3 + c^3} = \dfrac{a^3 + b^3 + c^3}{a^3 + b^3 + c^3} = 1.$

Therefore, we get: $a^2b^2c^2(a^3 + b^3 + c^3)^{-1}(a^{-3} + b^{-3} + c^{-3}) = 1.$

Quite often, we can put an expression in some different ways.

So the same expressions can look different.

And thus, we can put $a^{-3} + b^{-3} + c^{-3}$ the way below, too:

$$a^{-3} + b^{-3} + c^{-3} = \frac{b^3c^3 + a^3c^3 + a^3b^3}{a^3b^3c^3}.$$

Next, we can get: $a^2b^2c^2 = a^2acc^2 = a^3c^3$ since $b^2 = ac$.
So we get:

$$a^2b^2c^2(a^{-3} + b^{-3} + c^{-3}) = \frac{a^3c^3(b^3c^3 + a^3c^3 + a^3b^3)}{a^3b^3c^3} = \frac{b^3c^3 + a^3c^3 + a^3b^3}{b^3}$$

$$= \frac{b^3c^3 + b^6 + a^3b^3}{b^3} \quad \text{(since } b^2 = ac \Rightarrow b^6 = a^3c^3.\text{)}$$

$$= c^3 + b^3 + a^3.$$

Therefore, we get: $a^2b^2c^2(a^3 + b^3 + c^3)^{-1}(a^{-3} + b^{-3} + c^{-3}) = 1.$

In short:

We have: $a^2b^2c^2(a^3 + b^3 + c^3)^{-1}(a^{-3} + b^{-3} + c^{-3}) = \dfrac{a^2b^2c^2(a^{-3} + b^{-3} + c^{-3})}{a^3 + b^3 + c^3}.$

Meanwhile, $a^2b^2c^2(a^{-3} + b^{-3} + c^{-3}) = a^{-1}b^2c^2 + a^2b^{-1}c^2 + a^2b^2c^{-1}$.

So we get:

$b^2 = ac \Rightarrow a^{-1}b^2c^2 = a^{-1}(ac)c^2 = c^3$.

$b^2 = ac \Rightarrow b^4 = a^2c^2 \Rightarrow a^2b^{-1}c^2 = a^2c^2b^{-1} = b^4b^{-1} = b^3$.

$b^2 = ac \Rightarrow a = b^2c^{-1} \Rightarrow a^2b^2c^{-1} = a^2a = a^3$.

Thus, $a^2b^2c^2(a^3 + b^3 + c^3)^{-1}(a^{-3} + b^{-3} + c^{-3}) = a^2b^2c^2(a^{-3} + b^{-3} + c^{-3})(a^3 + b^3 + c^3)^{-1}$
$= (c^3 + b^3 + a^3)(a^3 + b^3 + c^3)^{-1} = 1$.

www.ingramcontent.com/pod-product-compliance
Lightning Source LLC
Chambersburg PA
CBHW081122170526
45165CB00008B/2522

* 9 7 8 1 4 6 6 4 3 8 1 7 0 *